Invisible Giant

Invisible Giant

Cargill and its Transnational Strategies

Second edition

Brewster Kneen

Pluto Press
LONDON • STERLING, VIRGINIA

First published 1995, this edition 2002 by Pluto Press
345 Archway Road, London N6 5AA
and 22883 Quicksilver Drive,
Sterling, VA 20166–2012, USA

www.plutobooks.com

British Library Cataloguing in Publication Data
A catalogue record for this book is available from the British Library

ISBN 0 7453 1959 9 hardback
ISBN 0 7453 1958 0 paperback

Library of Congress Cataloging in Publication Data
Kneen, Brewster.
  Invisible giant : Cargill and its transnational strategies / Brewster
Kneen.— 2nd ed.
      p. cm.
Includes bibliographical references and index.
  ISBN 0–7453–1959–9 (hbk) — ISBN 0–7453–1958–0 (pbk)
  1. Cargill, Inc.—History. 2. Grain trade—United States—History.
3. Agricultural industries—United States—History. 4. Food industry
and trade—United States—History. 5. International business
enterprises—United States—History. 6. Grain trade—History. 7.
Agricultural industries—History. 8. Food industry and trade—History.
I. Title.
  HD9039.C37 K58 2002
  338.8'873'0973   dc21

                                                               2002005038

10   9   8   7   6   5   4   3   2   1

Designed and produced for Pluto Press by
Chase Publishing Services, Fortescue, Sidmouth EX10 9QG
Typeset from disk by Stanford DTP Services, Towcester
Printed in the European Union by Antony Rowe, Chippenham, England

Contents

Preface

Invisible Giant Applies New Make-Up

February 20, 2002. I was making a routine check of Cargill's website to see what they had acquired the previous week. To my surprise, I was not greeted by a news release on their latest deal, but by 'the new Cargill' and its new logo. Gone is the old green toilet seat – the stylized green 'C' with the teardrop shaped centre with the fully detachable Cargill name. The new logo – with the upper three-eighths of the old green 'C' arching gracefully from the 'a' over to dot the 'i' in Cargill – integrates the Cargill name in the logo, signifying that Cargill has decided it will no longer be quite such an invisible giant.

It was almost exciting.

The new logo is intended to convey that Cargill is 'building on tradition while moving forward', and as you will find herein, Cargill has indeed been moving relentlessly 'forward'.

As Cargill explains:

> The green banner of colour in the new Cargill logo connects visually to our previous logo ... There is much we want to keep from the Cargill that grew organizationally and geographically during the 36 years the logo was used ... At the same time, a new Cargill is taking shape – one that is more approachable, innovative and forward-facing ... Our fundamental business purpose is about nourishment, growth and making connections – to harness our knowledge and energy in providing goods and services that are necessary for life, health and growth.

And of the new hype for the new Cargill:

> Cargill entered the 21st century on a new journey – to become, within the decade, the premier provider of solutions to our food and agricultural customers ... We bring to this journey our traditional strengths of integrity and dependability. We also bring our accumulated expertise across many geographies, products and

services. On this foundation we will build stronger relationships with each customer. That will enable us together:

- to explore unmet needs;
- to discover jointly best ways of filling them;
- to create unique and valued solutions;
- and to deliver them reliably.

Reading these words, carefully crafted and probably at considerable expense, I am struck by their subjectivity and lack of content. I am also struck by their emphasis on what seems to be called today 'values' – which are identified as 'dependability' and 'integrity'. Not bad values, actually, but by themselves they don't tell us anything about the company's vision for us and for the world. This book is intended to give some content to this vision by looking at where Cargill has come from and where it is going.

The imprecision of Cargill's words is clever. The company is not making itself liable for anything, so it cannot be held to any promises by either its fans or its critics. It has not established any specific concrete goals – such as doubling in size every five to seven years, as it once rashly declared, or increasing its earnings, or being the biggest this or that.

What may really be new about the new Cargill is an acknowledgement, in both philosophy and practice, that cooperation is superior to competition as a way of doing business, though today the term is 'partnership' or 'joint venture'. Gone is the mean old trader buying low and selling high. Well, not really gone – the company is still willing to take advantage of the misfortune, or mismanagement, of others by buying their facilities cheap, and the company is still a global trader in an almost endless list of commodities using its capital leverage to make deals that most mere mortals can seldom dream of. But then Cargill is not mortal. It is the essence of corporate being, exercising an immortality that the immoral engineers of biotechnology still only fantasize about.

Cargill apparently no longer seeks to take advantage of others, but to give them advice, as 'partners'. At least this is what it tells farmers and the purchasers of its specialized food components and ingredients.

'We are undertaking a fundamental change in our approach to doing business,' said company chairman Warren Staley. 'Our efforts today centre on creating distinctive value for our customers – from helping farm customers market their products to helping manage food customers' supply-chain logistics and risks. With this strategy we are more customer-focused, performance-oriented and innovative in all of our business relationships.'[1]

In offering advice, as the seller of 'inputs' and the buyer of 'product', what Cargill is really doing is creating agricultural policy from the bottom up. In helping the farmers to grow more of what Cargill requires as its 'inputs' for trading and processing, and helping farmers sell their 'product' in the global system over which Cargill exercises considerable control, Cargill is building the kind of industrial agricultural system it can best profit by, not necessarily the one that serves the farmers or the public best, or the system that ensures that everyone everywhere is adequately nourished.

This second edition of *Invisible Giant* was undertaken because of a renewed interest in the broad issue of increasing corporate concentration and control in every economic sector, but particularly in agribusiness. Since the first edition went to press at the end of 1994, I have exercised 'due diligence', dumping every bit of information pertaining to Cargill I came across or was sent into a file on my hard drive. Then I had to sort it all out and try to make sense of it. I've also done some more travelling.

I am sure that Cargill changed its information policy after *Invisible Giant* first appeared. It did not want *Invisible Giant* to be the only source of information about the company, so it began to make much greater use of the Internet to supply the public with news of its activities, but at the same time it has been very careful not to put so much information in public view that the novice could actually know what the company was up to. A trick it often pulls is to post information without a date, or to post information and then not update it. Many of its web pages are three years old or more, making the information interesting but not necessarily accurate. Some information is posted and then withdrawn – press releases, for example – leaving a little 'not available' sign posted where the news had been. One has to wonder what prompted the withdrawal. The company has also removed any corporate publications from the reception areas of their offices and facilities.

As a result of this, and my inability to travel the globe talking to local Cargill people as much as I would like – which has always been the best source of reliable information – the reader will find that not everything is as up to date as one would prefer. I, too, would like to know how this or that turned out, or what happened next. I certainly would like to know, for example, what Cargill – and Monsanto, or Cargill and Monsanto as Renessen – are doing in China.

Dartmouth College historian Wayne G. Broehl Jr was commissioned by Cargill to write an official history of Cargill from 1895 to 1960.* The resulting book, the only official history of the company that I am aware of, provides the kind of visibility for the corporation that Cargill obviously wanted. With over 1,000 pages, its thickness makes it highly visible on the bookshelf. It also contains many good stories (in voluminous detail) of Cargill encounters with financiers, government agencies and competitors, as well as stories about the internal dynamics and personalities of the company. In fact, there is so much detail that the whole becomes invisible – or *un*-visible. I don't think this is accidental. Duncan MacMillan also wrote a colourful and very personal two-volume illustrated family history** and I have drawn from these two books for the brief historical sketch of Cargill and for other bits of historical information.

While there is also a lot of detail in this book, it is organized and presented in a vastly different fashion, for a different purpose, than either of the books just mentioned. What I have attempted to construct, using the information available to me, is an outsider's guide to understanding how Cargill works, where it has been and where it is going. There are many stories and reports that are not included here – about price-fixing charges, environmental pollution, public relations at the community level and the kind of charitable activities that any responsible corporation would want to be engaged in as they displace governments in the affairs of the world. While these are important, they do not say much about corporate strategy.

Over the years, I have talked with many Cargill employees, competitors, academics and government personnel in many countries, and I have acquired as much of Cargill's own literature as I could find. I have also regularly read the trade journals cited as references and monitored many other business magazines. Friends and librarians who believe that information is for sharing have also been

---

* *Cargill – Trading the World's Grain*, University Press of New England, New Hampshire, USA, 1992.

** W. Duncan MacMillan, with Patricia Condon Johnson, *MacGhillemhaoil – An Account of My Family From Earliest Times*, privately printed at Wayzata, Minnesota, 1990 (two volumes, illustrated).

helpful in sending snippets and clippings. The whole project is very much like assembling a jig-saw puzzle. While there are still many pieces missing, you can get the picture, and I hope that others will use this as a starting point collectively to create a fuller picture.

I have endeavoured to make the statistics unobtrusive and helpful. There are many figures missing, for which the reader will have to apply directly to Cargill. US currency figures are used except where conversion from foreign figures would introduce error due to changes in exchange rates. Figures, however, are only one indicator; when the issue is power, the magnitude of the numbers themselves may be less important than the leverage they provide.

Cargill is certainly one of the most powerful and effective corporations in the world, and deserves to be known and understood. Cargill has and will continue to shape the agricultural policy of as many countries and regions as it can, while the public's role in this policy is confined to that of passive consumer. Public policy should be made by the public, however, and there are fundamental choices to be made about how we and future generations are going to live and how we are going to feed ourselves. These choices should not be left to Cargill or any other transnational corporation (TNC), regardless of the quality of its employees. It is simply not a good idea to put control over our food in the hands of a very small number of men whose job it is to serve corporate as opposed to public interests. There is a difference!

Here's to all those who do not share the Cargill vision of a single 'open' global food system or the mythical dogma of comparative advantage: the resisters in Venezuela, India, Argentina, Brazil, the USA, Canada and everywhere else who seek justice, equity, diversity and food for all.

1   Mutant Giants

Charles Darwin said, 'It is not the strongest of the species that survive, nor the most intelligent; it is the one that is most adaptable to change.' – Ruth Kimmelshue, Cargill, September 11, 2001

The authoritative trade journal-of-record for the 'grain-based foods' industry, *Milling & Baking News*, summed up the current situation of this sector of the food system in a blunt editorial in mid-2001:

> The international grain industry, the part of grain-based foods that at one time stood as the all-powerful core looming over all other sectors, has experienced dramatic upheaval in the past year or so. Not only has the power of the grain trade, once considered beyond challenge, been diminished in many different ways, but questions are properly being asked about the future of an industry that has been so radically transformed ... How business in grain is done, whether internationally or domestically, has been so hugely changed that it's appropriate to declare that the grain business ruling the 20th and 19th centuries is no more.[2]

When Dan Morgan wrote *Merchants of Grain* in 1979, there were five global grain companies with similar lines of businesses: Bunge, Dreyfus, André, Continental Grain and Cargill. André (founded in Switzerland in 1877) continues to operate on a much diminished scale after surviving bankruptcy proceedings; Continental Grain (founded by Jules and René Fribourg in France in 1921) recently sold its grain business to Cargill; Dreyfus has narrowed its traditional interests in grain trading and expanded its historic financial risk management function to be a provider of this 'service' to other companies around the world; and Bunge and Cargill are deeply into substantially reorienting their businesses, though Cargill remains the undisputed ruler of the grain trade while also extending its tentacles into every aspect of the global food system. Its 2001 corporate brochure describes 'Our Company':

> Cargill is an international marketer, processor and distributor of agricultural, food, financial and industrial products and services.

We provide distinctive customer solutions in supply chain management, food applications and health and nutrition.

We are the flour in your bread, the wheat in your noodles, the salt on your fries. We are the corn in your tortillas, the chocolate in your dessert, the sweetener in your soft drink. We are the oil in your salad dressing and the beef, pork or chicken you eat for dinner. We are the cotton in your clothing, the backing on your carpet and the fertilizer in your field.

Now that Cargill has purchased its grain handling business, what is left of Continental illustrates the specialization that has taken place. The company has been renamed ContiGroup Companies Inc. and consists of a cattle-feeding business (the largest in the US), an integrated pork-production business (the third largest in the US), the sixth largest poultry company, and animal feed and aquaculture businesses. At a time when the bottom has been driven out of the pig business, for the primary producers, that is, it should be noted that Conti acquired Premium Standard Farms (pigs) and Campbell Soup's poultry operations recently. The fastest growing segment of Conti's business, however, has been ContiFinancial, its commercial and consumer finance company. Perhaps ironically, this is an area in which Cargill has become much more cautious.

Of its purchase of Continental's grain business, Cargill said: 'We are going to create a much more competitive infrastructure to take grain off the farm and bring it to customers around the world. Producers will get a better price, and consumers will get a better price.'[3] The question farmers have to ask is, *why* should Cargill want the primary producer to get a better price? The question the public has to ask is, *why* should Cargill want the primary producer to get a better price? The major motivation for 'globalization' has not been to ensure that the primary producer gets a fair price, stays on the farm and feeds the family and the community, but to provide reliable and cheap access to raw materials, anywhere in the world, for the so-called 'value-adding' activities of the food-system giants.

The 'value' that is being added is not nutritional, however. It is shareholder value. Read any business paper to see what is valued and reported. It is not farm incomes, community economic stability, the health of the citizens, or equity and justice. The business news is all about returns to shareholders and record profits: 'ROYAL'S PROFIT IS BIGGEST IN CANADIAN HISTORY: Royal Bank of Canada made a stunning

$1.82 billion in fiscal 1998, the largest reported pure profit in Canadian history', the Toronto *Globe and Mail* proclaimed with pride on its front page. And just below it, 'Prairie grain farmers face devastation.' That Royal Bank profit works out to $60 for every man, woman and child in Canada the *Globe* told us – just about what pig farmers were losing on every pig they raised at the time.

Does Cargill – or Maple Leaf Foods or Smithfield – care, as long as there are farmers willing to go on purchasing pig feed from them or selling pigs to them at a loss so they can maintain shareholder value?

Cargill may be big, but Wal-Mart is bigger, and Cargill is happy to supply 'customer solutions':

> Wal-Mart's total sales in the year 2000 were $165 billion. Food sales accounted for 13% of that, or $22 billion. Increasing grocery sales are its top objective and growth vehicle. Cargill is a case-ready pioneer, ready to meet Wal-Mart's objective of handling only case-ready meats.[4]

I have seen no mention of Cargill's relations with Ahold, the giant Dutch food retailer that by the end of 2001 was doing $24 billion in supermarket business in the US and another $19 billion in food service.

While doing the research for the first edition of *Invisible Giant* in the early 1990s, I discovered that there had been virtually nothing of substance published about the grain trade and agribusiness in general after Morgan's book in 1979. Apparently the corporate sector did not like the negative publicity it had received throughout the 1960s and 1970s and made a concerted effort, with great success, to redirect public animosity toward the state. This was marked, or accompanied, by the ascendency of Ronald Reagan to the presidency of the USA in 1981. Corporations became the good guys and the state the villain. Critical analysis of how TNCs were reshaping the world disappeared. Now, in 2001, the global food system, from seed to supermarket, is in the hands of alarmingly few very large corporations whose primary commitment is to maximizing shareholder value.

This new, updated and substantially rewritten edition of *Invisible Giant* illustrates how the largest private company in the USA, if not the world, continues to mutate, always with the objective of expanding the control of its business interests and our food. Before proceeding with this story, however, it is worth noting the changes taking place in Bunge and Dreyfus, since they pop up here and there throughout the story.

Bunge

Bunge Ltd was established in 1818 as a grain-trading company in Amsterdam. In 1859 the company moved to Antwerp and in 1884 to Buenos Aries and then São Paulo, Brazil. In the fall of 2001 the company went public with an initial public offering (IPO) of stock, having moved its headquarters from Brazil to White Plains, New York, in preparation for the IPO. The focus of Bunge's business for the past century has been South America, but it is the largest soybean processor in both North and South America and the world's largest exporter of soybeans based on volume. It is a major corn and edible-oil processor, the largest wheat miller in Latin America and the largest integrated fertilizer producer in Brazil. In the year 2000, 35 per cent of Bunge's gross profits were made in fertilizer, 35 per cent in 'agribusiness', 26 per cent in food products, and 4 per cent in 'other', with net sales of $9.7 billion. Net sales for 2001 were $11.5 billion.

In 1998 Bunge operated in ten countries with more than 37,000 employees, had 30 crushing plants, 20 oil-refining plants, 10 hydrogenation plants, 13 packaging plants, one protein plant, more than 200 silos and elevators, three port facilities and 450 barges. Its holdings included:

- 58 per cent of food conglomerate Ceval Alimentos of Brazil, the largest soybean processor in Latin America
- 77 per cent of Serrana in Brazil, the largest fertilizer and phosphates producer in Latin America
- 60 per cent of Molinos Rio de la Plata, Argentina's largest producer and distributor of food products
- 68 per cent of Santista Alimentos of Brazil, an integrated manufacturer of consumer products, baked goods, and wheat flour
- 100 per cent of Bunge Australia
- 100 per cent of Gramoven, one of Venezuela's largest food companies, sold to Cargill in 1998.[5]

The president and chief executive officer of Bunge North America, John E. Klein, in an interview with *Milling & Baking News*, said that before his father Walter Klein took charge in 1959, the company was essentially a grain-trading house. Taking over from his father 15 years ago, John Klein decided to focus strategically on the Mississippi river and its tributaries and continue to reinforce the company's position as a major exporter out of the St Lawrence Seaway with its deep-water grain elevator in Quebec City. 'Today, we have more storage capacity along the Mississippi River and its tributaries than any of our major competitors.'[6]

For a time, Bunge and Continental Grain Co. were engaged in a joint venture for exports, but when it became clear that Continental would not remain an independent company, Bunge established a joint venture with Zen-Noh Grain Corp to export grain from their terminals in Louisiana on the Gulf of Mexico in 1998. The creation of Bunge Global Markets in 1999 was the company's strategic response to privatizations of government buying agencies around the world in the aftermath of the shift from planned to market economies.

Klein estimated the company's soybean market share in the USA at 17 per cent, in third place behind ADM and Cargill. Bunge got into corn dry milling in 1979 with the purchase of the Lauhoff Grain Co. soybean-processing facility at Danville, Illinois, which came with two corn dry mills. Bunge Milling is now the largest corn dry-miller in the USA and probably the world.[7] (For a description of dry and wet milling see page 72, Corn Processing.)

The repositioning and consolidation of the major grain companies, and their increasing collaboration to eliminate competition, is clearly marked by relations between Bunge and Cargill. In May, 1995, the two companies reached an agreement whereby Cargill would acquire Bunge's export grain elevator at Portland, Oregon and swap its river elevator at Osceola, Arkansas, for Bunge's river elevator at Price's Landing, Missouri. Earlier in 1995 Bunge had sold to Cargill 19 elevators in South Dakota, Minnesota, Colorado and Kansas so that Bunge could concentrate its grain operations along the Mississippi River and its tributaries. Bunge was willing to give up its Portland terminal because it did not have upriver origination facilities. Cargill, on the other hand, said the terminal would enhance its ability to serve North American farm customers as well as customers in the Pacific Rim. Later that same year Cargill acquired BEOCO Ltd, Bunge's oilseed crushing and refining operation in Liverpool, England. This enabled Cargill to move into the bottled and boxed hard-fats business in the UK for the first time. In December 1998, Cargill acquired the Bunge Venezuela milling business, Grandes Molinos de Venezuela SA (Gramoven). Included in the purchase was a flour mill, a pasta plant and an edible-oils plant, all near Caracas.

Dreyfus

Established in 1851, Louis Dreyfus & Cie has been owned and managed by members of the Louis Dreyfus family ever since. Current chairman is William Louis Dreyfus. The company is headquartered in Paris. 'Fifteen years ago we really were a grain company', Louis Dreyfus Corp executive vice-president Bruce Ritter told *Milling & Baking News*. 'Today we see ourselves as a risk management firm. We have diversified into coffee, sugar, rice, cotton, meats, citrus and energy ... We believe we can bring value to any market in the world that has risk. ... Industrial companies today face price risk, sovereign risk, client risk and quality and logistics risk.'[8]

In 1993 Dreyfus formed a joint venture with Archer Daniels Midland (ADM) under which ADM assumed operational control of 46 Dreyfus-owned US grain elevators. This made ADM the largest grain company in the US, measured on the basis of elevator capacity. Dreyfus retained its export elevator in Quebec City.

More recently the company has expanded its US export capacity by buying three elevators from ContiGroup, including a large export terminal in Beaumont, Texas. It also leased the very large Public Elevator in Houston, Texas, and took over Cargill's lease on Pier 86 in Seattle. (We'll come back to these stories in due course.) To complement its new export capacity, Dreyfus has established a series of partnerships with regional companies that source grain for it. In a contrary move, Dreyfus has undertaken a major expansion of its interior grain storage capacity in Canada.

In October 2001, Louis Dreyfus Corp. and Cargill formed a joint venture called CLD Pacific Grain LLC to combine operations in the Pacific north-west. The business will include ten grain elevators with a total capacity of 15.8 million bushels. Cargill leased its two export facilities in Portland, Oregon, to CLD, as well as its six facilities on the Columbia and Snake Rivers between Portland and Lewiston, while Dreyfus turned over its export facility in Portland and its river facility in Windust, Washington. The announcement of the venture boldly stated: 'Cargill and Louis Dreyfus will continue to compete aggressively against one another for the export business', according to *Milling & Baking News*, though this notice was not posted anywhere on Cargill's website to the best of my knowledge.[9]

Louis Dreyfus Canada struck an interesting deal in 2002 with Dow AgroSciences to market the specialty canola crops produced from Dow's Nexera seed. Actually, the seed comes from Dow subsidiary

Mycogen. Nexera, a non-genetically engineered (GE) canola with an especially healthy oil profile, is rapidly gaining a significant share of Canadian canola production (5 per cent in a bare three years) and Dreyfus is getting in on the ground floor of distribution of identity preserved (IP) specialty crops.

Cargill's World

Corporations operating beyond national boundaries are nothing new but, until the late 1950s or so, they were just that. Then for a time they were referred to as 'multinationals', a term that implies they are composed of, or represent the interests of, many nations. Nestlé and Unilever, Cargill and Mitsubishi, however, neither consist of nor represent many nations. While these collective personalities have to be incorporated under the laws of some land of convenience or tradition, they owe loyalty to no state or nation. They cannot function in the interests of any particular country precisely because they have to serve the interests of the corporate persona and its owners first. They live everywhere and nowhere in a world of markets.

Very few people are aware of Cargill's global activities, and even fewer could describe them, including (judging by the many I have talked with) most of Cargill's own employees. This is no accident. A picture of the whole would be disturbing to many people and would reveal the power of the corporation. Experience suggests it is better to remain largely invisible. Example: the casual visitor to the Hohenberg office in Memphis, Tennessee, would be hard pressed to know that one of the major worldwide cotton-trading companies is a Cargill subsidiary. Example: in many towns and cities the Cargill office is not where one might expect it, but rather in a nondescript office building outside the main business area where there is no indication of Cargill's presence except on the list of tenants in the lobby. More than once, when calling on Cargill executives, from Tokyo to Warsaw, I have been asked: 'How did you find this office?' Example: one comes across a yard full of gleaming tanker trucks bearing the name Transportation Services – but one has to inquire in the office to ascertain that this is a Cargill subsidiary.

The cloak of invisibility, however, takes other forms than being un-visible; being private, for example. Cargill Inc. has always been a privately owned corporation (it has never offered shares for public purchase) and, like a private person, under corporate law it is not required to reveal its personal affairs. No quarterly statements, no annual reports, no disclosures for a bond issue (though Cargill did do that – once). Cargill doesn't even have to be forthcoming to those who give it credit ratings for the sake of suppliers and bankers. In its 1994 Business Report on Cargill, Dun & Bradstreet gave the company the strongest credit rating it offers, and perhaps it has done this every other year as well, but Dun & Bradstreet told me: 'The co-operation we get from them is not exactly 100 per cent.'[10] Cargill provides them with the company audit, but 'for summarization only'. This pretty well sums up the challenge when trying to report on or analyse Cargill Inc. (or most other private companies) though, since the first edition of this book appeared in 1995, Cargill has been making public what it calls quarterly and annual financial reports. These would not pass muster as audited financial statements, but they are not intended to. Cargill reveals only what it feels is in the best interests of the corporation.

Prior to the days of satellite imaging, one required a fertile imagination to see the world in terms of water and 'geographies' rather than as states or continents. My own favourite picture of the world is a composite satellite photo-map of the world that displays topography, highlights water, and is devoid of superimposed political jurisdictions. No counties, no provinces, no states, no nations, no World Bank, no United Nations. This image of the world is Cargill's starting point, even though it studiously cultivates relations with political jurisdictions at every level, from mayors to presidents and prime ministers.

What does Cargill see from its satellite perspective? A relatively simple picture of the major growing areas of the world, and the water routes that can or might connect them to the major markets of the world. Thus in Brazil, what Cargill sees is rivers of soy; not rain forests and clear-cut jungle, but the great plain of the Mato Grosso and its potential for soybean production, if only the water routes to the sea can be made navigable. What it sees on the Indian sub-continent are two global resource areas: Punjab, for grain, and the plains of the south central area for corn and oilseeds.* The problem

---

* In 1993 the Government of India reversed its food policy and removed the constraints on the export of staple commodities such as wheat and rice that had been in place since the country's independence. 'It cited the possibilities for developing internationally competitive crop production in fertile areas of the country, such as the northern state of Punjab' (*Milling & Baking News*, May 4, 1993).

with Punjab is its lack of access to 'global water'. And so on around the world.

I am sure that this 'ecological' sensibility is one of the reasons for Cargill's continuing success. Instead of allowing its activities and interests to be defined by existing political orders and structures, such as states, governments and even trade agreements, Cargill has started with populations, geographies, regions and water.

Once its strategy is in place, Cargill works out the tactics required to deal with the appropriate political jurisdictions. In fact, Cargill appears to devote far more energy to establishing favourable national or regional business climates wherever it chooses to do business than it devotes to international trade agreements. Cargill has been developing its own internal global trading arrangements far longer than the World Bank and International Monetary Fund (IMF) have been around.

Coupled with Cargill's emphasis on geographies and sectors is the use of military terminology in discussion of strategy. 'Beachhead' is the key term and strategic concept that has been in use for at least the last decade.

> Cargill speaks of beachheads ... Historic product-line beachheads for the company have been hybrid seeds (primarily corn), commodity export marketing, and animal feed milling. The strategy has been: create the beachhead with inputs of capital, technology and a management nucleus; get the cash flow positive; re-invest the cash flow and expand the beachhead.[11]

Public Policy

Cargill has never, in all its history, been shy of telling governments, at any and every level, privately and publicly, what they should do. Sometimes this is dressed up in economic development terms, sometimes in humanitarian terms, and often just as naked self-interest. Cargill's website, under 'speeches', carries a number of what it obviously considers the corporation's key pronouncements on public policy. Whitney MacMillan, who retired as chairman of Cargill in 1995 after 18 years as president and then chairman of Cargill Inc., bringing to an end 85 years of MacMillan family leadership in the company, made a point of presenting the Cargill scenario for global food production and 'food security' on carefully selected occasions:

There is a mistaken belief that the greatest agricultural need in the developing world is to develop the capacity to grow food for local consumption. That is misguided ... Countries should produce what they produce best, and trade ... Subsistence agriculture ... encourages misuse of resources and damage to the environment.[12]

Cargill vice-president Robbin Johnson argued the same line:

Breaking the poverty cycle means shifting from subsistence agriculture to commercialized agriculture. Subsistence agriculture locks peasants out of income growth; it leaves populations outside the food-trading system and therefore more vulnerable to crop disasters, and it harms the environment through overuse of fragile land resources.[13]

It doesn't take much to see through this self-serving panacea for the hungry. Cargill is a supplier of inputs, a buyer, trader and processor of commodities, and a speculator throughout the entire system. The arch enemy of Cargill is subsistence agriculture, self-provisioning, self-reliance, or whatever you want to call the alternative to being incorporated into its growing global system of dependency. If you have or want a secure and adequate monetary income, and are among the diminishing numbers of industrial farmers, the Cargill way may be your way for now. For the majority of the world's population, the Cargill way of industrial commodity production is the way first to dependency and then to starvation for lack of buying power.

MacMillan, however, saw Cargill and its way of life as the solution to world hunger:

I believe this century has seen the emergence of [an] institution ... capable of playing a major role in attacking the world hunger problem, if allowed to do so. That institution is the modern global company. Companies like Cargill ... do things that go to the heart of our hunger problem. We bring goods and services needed by people that are fundamental to their well-being. We create markets that otherwise might not be available. We bring needed capital, and we transfer technology and expertise that adds to the efficiency of the marketplace, and we transfer the economic gains from that added efficiency to both the people we buy from and the people we sell to.[14]

In 1996 Cargill vice-president Robbin Johnson presented a clear statement of the Cargill ideology in the Asia Pacific Economic Cooperation (APEC) review, 'APEC and Building Global Food System'. Cargill's favourite term to describe its political objective is 'open food system'.

> Self-sufficiency ... is not a practical answer to Asia's growing food demand. Expanded trade is necessary to smooth out regional supply swings and harness the productivity of low-cost producers worldwide. By tapping into the natural advantages and techno-logical gains of efficient food producers, they can avoid the Malthusian dilemma ... Food security is often mistakenly translated into a demand for food self-sufficiency. It does not have to mean that each country produces all of its own basic foodstuffs. In fact, an open trading system has three incontestable advantages over self-sufficiency ... First, trade reduces the risks arising from crop shortfalls ... Second, trade lowers food costs by giving consumers access to efficient producers ... Third, trade raises incomes and improves diets through comparative advantage.[15]

Such 'truths' are presented as self-evident and without need of sub-stantiation or argument. References are never provided.

Whitney MacMillan's successor as chairman and CEO of Cargill, Ernest Micek, tidily summed up the Cargill dogma in 1998:

> We should start by sorting out what trade can and cannot do. Trade can deliver large, widely shared benefits: Export-oriented industries pay better and have more secure jobs; consumers get better deals; competition keeps businesses on their toes, and trade expansion generates overall economic growth. That benefit goes to all trading partners ... By allowing the law of comparative advantage to work, trade can promote efficiencies and create wealth that can help protect the environment and advance human rights.[16]

Micek was well positioned to be heard by the US Government. For example, in 1997 Micek was chairman of the Emergency Committee for American Trade, an organization of the head of 53 major US companies that 'support US trade policies to enhance American com-petitiveness in international markets'. A year later he was appointed to Clinton's President's Export Council. The Council was established

25 years ago to advise the president on trade policy and to promote US export expansion. 'We don't make policy,' said Micek, 'but it's a very high-powered group. The president, Congress and Cabinet secretaries certainly take note.'[17]

Micek retired as CEO a year ahead of schedule in 1999, but maintained his position as chairman (until 2000) to enable Warren Staley, who began his career with Cargill in 1969, to assume the presidency and implement a new strategic plan for the company. (Micek was president from 1994 to 1998, Staley was president from 1998 to 2000 and remains chairman of the board and CEO.) Gregory Page was elected president and chief operating officer of Cargill in April 2000, taking over the president's role from Warren Staley, now 57. Page, 50, started his career with Cargill as a 22-year-old. He has worked in the company's meat-processing, animal-nutrition and financial businesses in various locations in the US as well as in Singapore and Thailand.

A Cargill press release on the occasion of Staley's move to the president's position quoted him regarding his predecessor Micek: 'Ernie has led this company with courage and wisdom through some of the toughest times in our 134-year history, and at the same time he has inspired us to progress from an asset-intensive commodities company to a knowledge-based, solutions-oriented enterprise.' He went on to articulate a vision for Cargill's experts, 'serving more as consultants, digging into farmers' businesses and showing them how Cargill can help them farm more profitably and efficiently across product lines, from meat to grain'.

Staley issued a series of memos to senior Cargill managers in June and July, 1999, outlining his plans and conveying his instructions regarding roles and responsibilities. In describing the new Corporate Center he said: 'It will focus its attention and efforts on setting and enforcing corporate strategy, providing resources, developing and coaching talent, and understanding and managing external trends and constituencies essential to our success. It will also be the steward of Cargill's culture, values and basic beliefs.' In elaborating on these basics, 'providing corporate resources' includes 'sharing and leveraging of resources (capital, people, knowledge) and establish boundary conditions for participation'. (This is clearly a 'learning' from Cargill's experience in speculating in the currency market, particularly the Russian rouble, which cost it several hundred millions of dollars.) 'Promoting corporate behaviours' includes 'Define, model and reinforce desired behaviours', while 'Managing constituencies' includes 'Pursue a public policy environment that facilitates Cargill's business success.' This was spelled out further regarding the role of regional directors, whose responsibilities include, 'Represent Cargill to senior political figures.'[18]

In a subsequent memo on 'Strategic Intent: Our commitment to grow', Staley wrote: 'We want Cargill to grow in ways that make us the premier customer solutions company in the agrifood chain.' He gave his managers the maxim: 'We must go to sleep thinking about growth today, dream about growth tomorrow, and wake up with insights for growth in the future.'[19]

This future orientation of the 'new Cargill' is reflected in the general age of Cargill's leadership, all under 60, and in the rapidity of leadership turnover in the past few years among men who have spent all their working lives with Cargill. This speaks volumes about the internal process of leadership development that certainly distinguishes Cargill from many other companies.

The Cargill outlook was expressed in a more recent speech by Cargill executive Jim Prokopanko, but apparently it was a little too candid. It was posted on the company website, but removed sometime prior to December 2001. This is consistent with what I have noticed in tracking Cargill for many years: the company engages in a continual process of revising and editing its own public material. Press releases are posted and disappear, or are blanked out with a 'not available' notice, making it necessary to continually monitor the company's new stories so that one can catch them 'on the fly' as it were. Prokopanko's talk provided a clear description of the company's shifting strategy in somewhat unusual terms – for an insider (which may be why his speech disappeared).

> Cargill has moved from simply trading commodities between countries to creating food products and food distribution systems in much the same way as television or automobiles are manufactured. Here's a real life tangible example. Cargill produces phosphate fertilizer in Tampa, Florida. We use that fertilizer in the United States and Argentina to grow our soybeans. Soybeans are then processed into meal and oil. The meal is shipped to Thailand to feed chickens, which are processed, cooked and packaged so they can be sent back to supermarkets in Japan and also to Europe. It's pretty complicated.[20]

Now, it seems, just as the industrial world of the west has failed to reduce hunger either at home or anywhere else, Cargill has curtailed this lofty vision of feeding the world to settle for the more mundane task of providing 'customer solutions' to the affluent of the 'developed' world.

Cargill's Ruth Kimmelshue spoke about this at a Pew Initiatives conference in 2001:

> For at least the last 10 years, Cargill has provided to customers, who were willing to pay, specific Identity Preserved products – white corn and food grade yellow corn for tortillas and snack foods; soybeans with desired qualities ideal for the manufacture of tofu. Similarly, some of our customers in Japan place significant value on corn varieties that are not treated with chemicals after harvest ... So, is identity preservation costly? Yes. But looking back, we would argue that not having an effective identity preservation system is an even more costly prospect.

I have often marvelled at Cargill's durability and ability to change and adapt while retaining a firm grip on its far-flung empire. The strength of its 'corporate culture' is obviously at least in part responsible for this, and it is a culture that its employees learn well. As reflected in the speeches quoted above, different people at different times have an uncanny way of delivering the same message and even using identical language. One can be forgiven for thinking the play is well scripted by a very small committee.

The theme for Cargill's current corporate advertising campaign, 'nourishing potential', reflects its altruism of service to others, even though the others are more likely to be Cargill customers, not the hungry.

> The phrase 'nourishing potential' has several levels of meaning. At its most philosophical, it refers to Cargill's corporate vision of raising the living standards and achievement potential of people around the world through efficient distribution of affordable food. At a more nuts-and-bolts level, we seek to work closely with our customers – food companies, for example – to help them achieve their fullest potential in business ... The customer is not looking for an ingredient, but for a final product with a desired mouth feel, nutritional value or ease of preparation. In this approach, the ingredient supplier must work closely with customers to know

what ends they and their customers, the consumers, seek – and together we discover the means to get there ... Our ultimate goal? To be the best ally any food company can have. To deliver products and services that help our food customers deliver a better sensory experience to the consumer.[21]

2   Cargill Inc. – The Numbers

Just a bunch of people trying to make a living – a family company.
– Barbara Isman, vice-president, Cargill Canada, 1994

Cargill is a private US company, established in 1865. In its 1997 'financial report' Cargill described itself as:

an international marketer, processor and distributor of agricultural, food, financial and industrial products with some 79,000 employees in more than 1,000 locations in 72 countries and with business activities in 100 more. Its trading and processing businesses include grains and oilseeds, fruit juices, tropical commodities and fibers, meats and eggs, salt and petroleum, as well as the production and sale of livestock feeds, fertilizers and seeds. Industrial activities include steel recycling and manufacturing and steel-related trading and processing. Financial businesses include financial instruments trading, distressed asset investments, structured finance, futures brokerage and leasing.

Cargill's global operations are directed via satellite and dedicated fibre-optic cable systems from its headquarters in Minnetonka, Minnesota, a suburb of Minneapolis.

There is no way to dispute the figures that Cargill supplies to the nosy public. There is no legal requirement for a private company to publish an audited financial statement, it has no public shareholders to answer to, and a corporation with as many divisions, subsidiaries, joint ventures and partnerships as Cargill has can hide profits and losses in a multitude of ways. Figures, however, do express relativities and scale.

In 1971 (fiscal year ended May 31), on the eve of Russia's massive grain purchases, Cargill reported that it had annual revenues (sales) of $2 billion. By 1982 this figure had jumped to $29 billion and by 1994 it had nearly doubled, to $47.1 billion. That year the company had estimated operating profits of $1.5 billion and net profits of $571 million, in spite of a $100 million loss incurred early in 1994 by its Financial Markets Division in the trading of derivatives (contracts based on mortgage-backed securities).[22]

In 1994 a Minneapolis newspaper published figures, obviously provided by Cargill, on the changes in Cargill's business activities from 1970 to 1990. They showed merchandising (trading in bulk commodities) dropping from 37.3 per cent of Cargill's business, as a percentage of net worth, to 17.6 per cent. Non-merchandising (processing of oil seeds, corn and flour milling; agricultural products, such as poultry, feed and seed; industrial products such as steel, fertilizers and salt; and financial services) increased from 62.7 per cent to 82.4 per cent. The only non-merchandising activity to show a decline was transportation, from 6.3 per cent to 2.3 per cent.[23]

Cargill's global revenues peaked in 1996 at $56 billion and held at that level in 1997.

Cargill's chief executive officer Ernest Micek told the *Wall Street Journal* in 1997 that Cargill controls 25% of America's grain exports, 25% of the oilseed crushing capacity, 20% of the nation's corn-milling capacity, slaughters 20% of the cattle and owns 300 grain elevators.[24]

For 1998, revenue dropped to $51 billion, and cash flow was down 15 per cent to $1.6 billion for the year. Cargill cited global excess capacity in grain handling and processing, the drop-off in Asian demand caused by economic turmoil, greater market unpredictability caused by El Niño and losses in Cargill's consumer finance businesses, which the company has exited. Cargill invested $1.4 billion during the year on acquisitions such as oilseed processing in South America and Europe, and building feed mills in Poland and China.

Earlier in the year Cargill received its second warning in two months from Wall Street rating agencies, which were concerned about the international grain giant's sharp decline in profitability and its risky financial division. (Such warnings are only significant as they might affect the interest rate a bank might charge for a private loan to Cargill.) Though a strong contributor to earnings, the financial division's rapid and diverse expansion also led to problems, highlighted by Cargill's $90 million charge to cover losses in a mobile-home lending business. A confidential document showed that in 1998 commodity trading and processing generated 93 per cent of Cargill's total revenue of $51.4 billion and that 45 per cent of its earnings were generated outside the US.[25]

In 1999, Cargill's reported revenues dropped again, to $46 billion, but began to rise again in 2000, with revenues of $48 billion. For 2001, with revenues of $49.4 billion, the company reported that it had made good progress in executing its corporate strategy 'to become a premier provider of customer solutions in food and agriculture'. Sectors performing well were meat processing, financial businesses, global grain and oilseeds, salt, and global petroleum and ocean transportation. Industry overcapacity hurt results in flour, juice and steel, and the continued weakness in agricultural markets was hard on farm services and fertilizer production, the company reported.

	1971	1982	1994	1996	1997	1998	1999	2000	2001
Revenue totals ($billion)	2.0	29.0	47.1	56.0	56.0	51.4	46.0	48.0	49.4

In January, 2002, Cargill reported earnings of $522 million for the first six months of fiscal 2002, a 51 per cent increase from the net income reported in last year's first half. 'Our results reflect a journey Cargill began several years ago to be less dependent on buying and selling commodities and more invested in solving our customers' problems', said Warren Staley, chairman and CEO. 'We're doing more to help farm customers market their output successfully and to help food customers manage their supply-chain risks. That enabled us to post strong earnings in spite of the tragic events of September 11, the deepening financial crisis in Argentina, the sudden bankruptcy of a major energy trader [Enron] and a weak global economy.' Looking ahead, Staley said: 'At our core, Cargill is about nourishing people. We're in a knowledge business, and we are changing how we come together as a company of diverse and enterprising people to make a valuable difference for food and farm customers.'[26]

Cargill now describes itself in its press releases (or at least it did before February 20, 2002) as: 'an international marketer, processor and distributor of agricultural, food, financial and industrial products and services with 90,000 employees in 57 countries. The company provides distinctive customer solutions in supply chain management, food applications and health and nutrition.' Cargill no longer tells the public the number of 'locations' in which it does business. Perhaps it wants to play down just how extensive its operations are. Employing a whole lot of people, on the other hand, is a good thing.

The company also seems to place great importance on its identity as defined 'by the company we keep'. The company, identified as 'customers and partners', are the likes of McDonald's, Kraft, Nestlé, Coca-Cola, PepsiCo, Kikkomen, Wal-Mart and Unilever. Which does raise some questions about 'nourishing people'.

3 Origins, Organization and Ownership

The official histories give 1865 as the founding year for Cargill Inc., but Duncan MacMillan, in his family history, writes that it was in 1867 that W.W. Cargill started in business for himself in Iowa and two years later that he and his brother Sam formed W.W. Cargill and Brother. They built grain elevators along the railways in Minnesota and Wisconsin and after the financial panic of 1873 the brothers took advantage of the opportunity to buy up properties cheaply, a practice that has endured to this day. They entered partnerships with others, a pragmatic practice that the company continues to exploit, to trade in commodities other than grain, such as wool and pigs, and they traded in chickens 'by the carload'.

Cargill and partners were soon also buying land, and, as Duncan MacMillan tells it, by the end of 1879 the Cargill & Van partnership farm had become a small village, with 1,000 sheep, 300 pigs, horses and small stock. W.W. Cargill also began experimenting with seed breeding. This was half a century before the 'invention' of modern hybrid corn.

By 1881 the Cargill brothers, Will, Sam and Jim, were in business as Cargill Brothers and the partnerships with outsiders were ended. They began a period of rapid expansion to the north up the Red River Valley and to the north-west. In 1884 the Cargills moved their office to Minneapolis, which was fast becoming the grain-milling centre of the country.

LaCross, Wisconsin, was home for the Cargill family for a time, and just across the road lived Duncan McMillan and his family. The McMillan clan had made their initial fortune logging and lumber milling. In 1895 Edna, eldest daughter of W.W. Cargill, married John MacMillan, son of Duncan McMillan (the spelling of the family name changed at that point).

The last years of the nineteenth century and the first decade of the twentieth were tumultuous and nearly disastrous for the Cargill businesses. Will Cargill, son of W.W., had become involved in some ambitious land development schemes in Montana that syphoned funds through his hands without the full realization of the families, leaving the entire dynasty on the verge of bankruptcy. The truth came to light in 1909 when Will Cargill died rather suddenly. In classic manner, the creditors descended, hoping to recover at least some of what they had invested or loaned. Most of the assets, however, were not liquid, leaving creditors with the choice of sticking it out or forcing the liquidation of the Cargill companies. Fortunately for the Cargills, John MacMillan Sr, who was already deeply involved in the businesses, had just extended his personal credit resources and was able to convince the creditors that it would be better in the long run for them to stay with the company, under his leadership, than to force its collapse.

The history of the company since then has been one of relentless, though at times jerky, growth. Mistakes have been made, but Cargill as a company has developed great skill at knowing when to make a strategic retreat and when to stand fast.

In the late 1930s when the grain fields turned into the Dust Bowl, the company bought up all the corn futures on the Chicago Board of Trade (CBOT) and was charged by Secretary of Agriculture Henry Wallace with trying to corner the market. Cargill pleaded innocent, of course, but was nevertheless suspended from the CBOT. The affair dragged on for three years until 1940: 'when Cargill was allowed to plead not guilty, in return for the denial of the trading privileges on the CBOT of Cargill Grain Co of Illinois and of John H. MacMillan. Everyone knew that MacMillan did not trade personally and that the Cargill Grain Co. of Illinois was being liquidated.'[27] In other words, a deal was made and business carried on as usual.

Cargill Inc. fared well during World War II, enjoying government contracts for ship building and grain storage and delivery, and serving the needs of agriculture. 'The Chase National Bank ... complimented the Company in the spring of 1945 on its strong financial position: "The working capital has been more than doubled since May of 1941 – a real accomplishment from which you should derive much satisfaction."'[28]

W.G. Broehl makes it abundantly clear that there have been, at times, difficult internal debates over corporate structure and leadership. This became a major preoccupation again at the end of the turbulent 1980s as it became clear that the company would have to be restructured in order to stay on top of its diverse and evolving worldwide business activities and that the ownership structure would have to be altered to accommodate the diverging interests of the family owners.

Restructuring

The North American Organization Project (NAOP) was formed in 1990 after Whitney MacMillan had convened a group of 50 managers to talk about their corporate vision and their strategy for the future. They also had to decide how to bring Cargill's North American businesses into line with the way the company's other worldwide businesses were managed – what they called a 'soft matrix' of both 'product line and geography management'. The outcome, two years later, was a decision to form the Corporate Centre, composed of senior management, to concentrate on strategy, asset allocation, and personnel decisions, leaving operating decisions to the next level of management, the 'geographies'.[29]

The restructuring was motivated by the realization that personal relations, and information-sharing and decision-making capacities were all at risk of being swamped by the company's growth and diversity. 'Twenty years ago most of the decisions in this corporation were made down at the coffee table', explained Cargill president Heinz Hutter. 'People met every day there and met every day for lunch in the dining room. That's the way information was shared, and everyone knew what was going on. Well, we're too big for that now.'[30]

The basic enterprise that Cargill began with, and pretty much stuck with until the 1930s, was trading, transporting and storing grain and other commodities, commodities being really any bulk material that could be handled as dry generic (undifferentiated) cargo. As it began to move upstream into seeds, fertilizers and feeds, and downstream into milling and processing, and then into a broader and broader range of commodities and products, the centralized structure of the organization was constantly under stress.

It must have been the magnitude of the 1992 restructuring that caused Cargill to have *Milling & Baking News* carry an unusually long story about it. I think it likely that Cargill felt it had to reassure its many customers and competitors that it was not restructuring as a consequence of some form of ill-health or mis-management, but to continue to perform according to its stated goal of doubling the size of the corporation, meaning its equity, every five to seven years. (The company's equity was valued at $4.1 billion in 1992 according to the interpretation of the shares tendered by the family members at the time.)

The major aspect of the restructuring was the designation of North America (including Canada and the US, at least, although neither

Canada nor Mexico are ever mentioned) as a 'geography' on a par with the other geographies of Europe, South America, south-east Asia and north Asia. These are the principal regions within which the company does business, but the business is carried out along product lines as well. The product line and the geography together determine the 'matrix' of activity.

Of course, it is not that simple, because designating North America as a geography carried with it the creation of the Corporate Centre with its responsibility to establish the vision and goals for Cargill on a worldwide basis. 'We want to do less controlling and more strategizing about future directions', said Whitney MacMillan.[31] He added that Cargill had been 'too inclined to want to measure everything, when we ought to look at something as a continuous belt without focussing on each minute step'.

Interestingly, Cargill's arch competitor ADM uses similar imagery to describe how corn processing yields 'an enormous river of dextrose' that flows between the farm and the final products. ADM chairman Dwayne Andreas also described his very complex company as a 'single business' with the result that 'the increasing complexity of operations makes it more difficult than ever to accurately separate profits and losses of various raw materials from one another. We make arbitrary decisions ... Many costs ... are arbitrarily allocated.'[32]

In other words, ADM, like Cargill, had reached a stage where it was no longer reasonable to speak of isolated 'profit centres' or even 'product lines' and where corporate welfare as a whole supercedes the accounting of its parts. There is here an expression of organic complexity and the biological phenomenon of the whole organism being more than the sum of its parts. This is particularly notable in contrast to the reductionist ideology driving agricultural biotechnology.

To end the confusion between Cargill the global corporation and Cargill the North American business (which accounts for more than half the corporation's net worth), the company made another structural change, creating a new Cargill Grain Division (CGD) to replace the old Commodity Marketing Division (CMD). The CGD is now the primary source of grains and oilseeds in North America for both domestic and export customers, including its own corn milling, flour milling and oilseed processing. In practice, this means that instead of various Cargill locations competing for the same grains or oilseeds, CGD will make a single bid for each commodity in specific originating locations. This will put an end to 'transfer pricing' between competing Cargill entities. 'The market will rule' is the way this is described.

Rather than acknowledge this as centralized monopoly bidding, the company interprets it to mean that 'merchandisers and product line managers at decentralized locations will be empowered to purchase grain when it is offered at market prices'.[33] Cargill also contends that suppliers (farmers and other) find advantages in dealing with only one buyer who bids 'at the market'.

Cargill may think of this as empowering local managers, but it also explains why local managers have two satellite dishes: one to receive prices from the Chicago futures market, the other to receive orders from Cargill headquarters in Minneapolis. The local managers may be held responsible for their decisions on buying and selling, but the rules they play by are clearly determined moment by moment by the central authority. Marx would have loved the contradiction.

In effect, Cargill returned to the centralized structure it had before the great explosion of grain trading in 1972–3 and the period of decentralization that followed. Perhaps in the 1970s and 1980s it was simply not possible technically (or technologically) to centralize the burgeoning trade in commodities, but now electronic communications make it a relatively simple matter.

It is, however, more than that. The number of large trading companies has been greatly diminished and the power of the remaining few magnified accordingly. There are no longer many buyers and sellers, and the few left do not really wish to compete, either intentionally or by accident; Cargill's bid may be the only one at a particular location on any particular day; and a single buyer for a large quantity can exercise considerable leverage on the market. Pity those who believe the propaganda of 'competition'.

The CMD was not the only piece of the company to be restructured. Cargill's worldwide marketing of all meat and fish products was brought together in a single business unit and is now handled in the same way as oilseed processing and corn milling. This makes their slogan about global sourcing into a concrete reality. Cargill does not simply have plant A ship x amount of its product to customer B. If the product of the moment is manufacturing beef for Canadian hamburger fabricators, it might come from a Cargill plant in Australia, the US or Canada itself, depending on local market conditions, what's in the cooler, and transportation logistics at the time.

The new 34-person Corporate Centre, with its responsibility to provide vision and leadership for the company's worldwide operations and corporate strategy, consists of 'senior line managers and five core corporate functions, plus support staff'. One aspect of its mission is to 'manage key constituencies to help shape an environment in which Cargill can prosper'.[34]

When interviewed by *Milling & Baking News* about the company restructuring, CEO Whitney MacMillan stated: 'There are just too many opportunities within our existing competencies to see any need for a significant change in [corporate] direction.' He said there were three paths that could take the company in its chosen direction: transferring existing competencies to geographies where the company is not currently active; finding other businesses where existing competencies could be applied and where Cargill has an advantage; and moving up the food chain 'at the margin'.[35]

MacMillan used seed corn as an example of the company going outside its core competencies (many years ago at a time when it was strictly a grain trader) to become the leader outside the US (apparently suggesting that Cargill's sales of seed corn outside of North America surpass those of industry giant Pioneer Hi-bred). It did this, he said, because of seed corn's unique role in providing 'a beachhead to enter developing countries where we might have been unwilling to operate under normal circumstances but where, with seed corn, we can establish a presence'.[36] Eight years later, in 1998, Cargill sold its landing craft, its global seed business (outside North America) to Monsanto, but kept the beachhead strategy.

As part of its global restructuring, and an indication of new directions for corporate expansion, in 1994 Cargill created a Specialty Plants Products Department for customers who have particular requirements, such as what are now being referred to as IP crops that have special characteristics specified by the end user. These crops include popcorn, organic grains, grains with specific baking characteristics, and oilseeds yielding an oil with particular cooking qualities. Some of these specialty crops are being created through genetic engineering.

These crops are obviously grown from special seed, but that is only the beginning. They have to be identified and kept isolated from planting through harvest and delivery to the end user, including any processing required. In a way, handling these crops is very much like handling grain in the days before the futures market and the commodity exchange, when grain was bought and sold on the basis of actual samples. In the case of IP crops, speculation is excluded, or held to very narrow margins, by the nature of the commodity and its ownership. In some cases, the end user, such as Procter & Gamble, specifies the characteristics desired, in a cooking oil, for example, and contracts with a seed company to create (genetically engineer) an oilseed that produces such a product. The seed company then contracts with farmers to multiply the new seed variety to produce enough seed for commercial production of the crop. Then the seed company sells the seed to farmers who grow the crop under contract to the processor that requested the oilseed in the first place. The end-user thus owns the seed and its product from beginning to end.

InnovaSure

In 2000, Cargill formalized its IP business in corn as 'InnovaSure', and the best indicator of Cargill's understanding of where global food is heading is found on a dedicated Cargill website, www.innovasure.com, which opens with the following:

> Dynamic changes are redefining the food industry. Consumers are paying more attention to what's inside the foods they eat. And in many countries, regulators are calling for more information on food labels. To succeed in this evolving marketplace, you need a supplier who understands the new issues you face. A supplier who can help you satisfy the demands of both consumers and regulators.

The various pages of the website then explain the InnovaSure (*innova*tion + as*sur*ance) IP system Cargill has built around its subsidiary, Illinois Cereal Mills. (The ads for InnovaSure represent it as a joint venture between Cargill and Illinois Cereal Mills, with no indication that Illinois Cereal Mills is actually a wholly owned Cargill subsidiary.)

> We go to excruciating lengths to ensure that the identity of our corn products remains intact from the time the seed is selected, until InnovaSure products arrive at your door ... We have been perfecting our fully traceable, fully documented systems for many years; leveraging world-class technology to bring you identity preserved products that help you to succeed with your customers.

Traceability is the cornerstone of our identity preservation system. We have the most stringent IP protocols and traceability systems in the industry ... All documents, samples and test results are retained for a minimum of two years and are available for third-party inspection.

Only non-genetically enhanced (conventionally bred) varieties are included on the approved InnovaSure hybrid list ... InnovaSure seed suppliers must be able to ensure the integrity of their products is maintained and that they have complete traceability throughout their system.

We partner with more than 400 professional growers ... Growers document that fields designated for IP corn have been free of genetically enhanced corn for at least one year. Growers identify the corn hybrids planted by their neighbours. They maintain proper buffer zones and submit a map showing what hybrids are planted in surrounding fields.

Ownership

As already mentioned, a change in ownership structure accompanied the internal corporate restructuring, or perhaps it was the other way around, with the need for changing the ownership structure seen as a good opportunity to reorganize.

In 1986 it was reported that Cargill's $2.6 billion in shareholder equity was held by fewer than 50 descendants of the Cargills and MacMillans, and 450 others, all current members of the company's management, who received about $10 million a year in dividends.[37] In 1992 *Fortune* magazine reported net worth (or shareholders equity) of the company as $3.6 billion as a result of 'a compound annual rate of growth of 12.2 per cent over the past 50 years'.[38]

Such a concentration of ownership not only results in huge rewards to the very limited number of owners, but also creates a very unusual kind of problem: what are the family members to do with all this money? The logical, and indeed capitalist, response is 'Invest it.' This is the basic reason that Cargill can boast, as it does in almost every country in which it does business, that it reinvests its profits in the country in which it makes them. (I think it actually does essentially this, but there is absolutely no way of knowing the real truth of the matter, as I have indicated elsewhere.) In fact,

Cargill says that since 1981 it has reinvested 87 per cent of its cash flow, with only 3 per cent of the company's profits being paid out in dividends. That kind of liquidity provides an awful lot of leverage, particularly when coupled with Cargill's somewhat unusual habit of using its capital to expand or enter into depressed commodity businesses.

This structure also created a problem. What would happen if a member of the Cargill–MacMillan clan decided that they would like to withdraw their Cargill 'investment' to apply it elsewhere – such as in a new radio station? There was no mechanism in place whereby they could receive cash instead of more equity in the company. When senior executive Dwayne Andreas left in the 1950s, his interest had to be bought out by the company, but there was no formal way of doing this and no book value to go by. (In Andreas's case, he actually did have shares in a Cargill subsidiary which could be redeemed, though at a somewhat arbitrary price.)

Cargill finally dealt with this by hiring a team of consultants to come up with a form of Employee Share Ownership Plan (ESOP) for the company that would ensure the security of corporate information; would ensure that the corporation would or could never go public; and would provide an escape or equity redemption programme for family members and, subsequently, senior managers. In 1992–3 Cargill successfully initiated the resulting plan by inviting family members to offer their 'shares' of the company for redemption. As it turned out: 'Family members tendered 17 per cent of their shares for $730.5 million, or an average of $8.3 million per family member.'[39] Reports at the time indicated that: 'Ownership ranks rose from fewer than 90 family members to nearly 20,000', but this is misleading because while the family members turned over ownership rights to 17 per cent of the company, this equity remained in the hands of the board of directors, held in trust for employees in the ESOP. At the time, 7,800 hourly employees and 12,000 salaried employees were eligible to participate in the ESOP.

Fortune magazine estimated the wealth of the three MacMillan family members at $2.1 billion, at the time, representing 8 per cent of Cargill Inc. (Cargill MacMillan Jr, in 2002 aged 73; Whitney MacMillan, 70; and Pauline MacMillan Keinath, 65).[40]

Control of Cargill, which is still believed to be the world's largest private company, is split among three branches of the Cargill and MacMillan families, according to Whitney MacMillan in a rare interview with Cargill's hometown newspaper, the Minneapolis *Star*

Tribune, in 1995. One hundred family heirs to the Cargill fortune control 83–85 per cent of the company's stock, with the remainder held by the company's ESOP on behalf of Cargill employees and by senior management. In other words, the family still controlled the company. When he assumed the presidency of Cargill in 1999, Warren Staley still had to keep the fourth- and fifth-generation Cargill heirs content with their dividend cheques. A confidential document that Cargill had circulated to a small group of investors interested in buying $250 million ten-year notes disclosed that Cargill had offered to use $106 million from the sale of its equipment-leasing business to repurchase shares from the descendants of the founding families.[41]

In 2001, Cargill asked its shareholders to approve a five-for-one split of its common stock. The move was designed to make it easier for owners to cash in small portions of their holdings. A financial advisor at the time appraised the fair value of a Cargill share at about $195, making the post-split value of a common share about $39. Some stock is held by a company-based stock-ownership plan for employees in which 22,000 workers participate. The last Cargill stock split was 16 years ago. The company has no plans to offer shares to the public.[42]

4   Policy Advocacy and Capitalist Subsidies

For decades prior to 1993 the monthly *Cargill Bulletin* was the sole public report of the corporation about its activities apart from limited news releases. The end of this era of keeping the public at least minimally informed was marked with the freshly designed January, 1993, issue. Instead of carefully selected information on corporate activities and developments, the *Bulletin* became a more academic presentation of global agricultural policy options, as defined by Cargill of course, along with a 'Cargill Commentary' providing the company's policy recommendations on the subject of each issue. The *Bulletin* faded away a few years ago, replaced by a careful selection of speeches posted on the corporate website, along with the other kinds of information I refer to throughout this text.

Policy advocacy is often put forward through the use of academic and professional policy analysts who can be hired to research, and present as objective, the policies that their clients wish to have implemented. The corporations paying the bills can then cite these 'independent' studies in support of their policy recommendations. It works, too, since the media, sharing the same corporate culture and ownership, go along with the game and quote the studies as objective and neutral, and without identifying the studies' sponsors even if they thought to ask about them. Cargill has made extensive use of this mechanism to further its policy interests.

It used to be common practice for corporations to use the revolving door of public service – ministries and regulatory agencies – to maximize their policy efforts and this is certainly still true, particularly in the drug and biotechnology area. Cargill, apparently, feels it is no longer necessary to use this approach, though senior Cargill management names frequently appear in various government and industry trade advisory and negotiating bodies. A good example occurred in the closing months of 1993, as the push was on to conclude the Uruguay Round of the General Agreement on Tariffs and Trade (GATT) negotiations, with the naming of the chief executives of two of the country's largest agriculture-related companies, H.D. Cleberg, of Farmland Industries, and Whitney

MacMillan, to a GATT advisory group 'to help congressional leaders monitor the final phase' of the GATT negotiations.[43]

Lobbying can also be effectively carried out at the grass-roots, as Cargill knows very well. The Cargill Community Network (CCN) is the name of a grass-roots programme 'aimed at improving Cargill's reputation and success in communities where it is doing business'. The CCN is designed 'to help win Cargill's public-policy objectives at every level of government' by spreading the word that Cargill is 'a solid corporate citizen' while 'building a reservoir of community goodwill that ensures we have friends when we need them'.[44] Whether or not Cargill still uses the name, the programme clearly remains in place.

The Ohio Circle is one example of a grass-roots campaign that achieved the results Cargill wanted, which was to defeat a statewide 1992 'right to know' initiative ('Issue 5') that would have given consumers and neighbourhoods in the state of Ohio more information about toxic substances used in the state. Polls showed Ohioans supported the measure by a margin of nearly nine to one, according to Cargill's own account. 'Because Issue 5 was a product of a grass-roots movement, it required a grass-roots movement to defeat it. That effort took shape in a group called Ohioans for Responsible Health Information.'[45] (Organizing citizens advocacy groups is a favourite tactic of the drug and chemical industries.)

First of all, Cargill managers from 20 locations met together for a 'Circle Meeting'. (Circle Meetings had been initiated by Cargill chairman Whitney MacMillan in 1984 as special gatherings of senior Cargill managers to learn what's new at corporate and local levels and to share ideas and information.) The first meeting produced the Ohio Circle Council which organized asset management, marketing, origination and public policy subgroups. The public policy group set about organizing a campaign 'to educate the voters of Ohio' about Issue 5. Cargill used its sales channels and trade associations to carry its message that the proposal would hurt Ohio businesses and the economy. 'We as a coalition of Cargill businesses in Ohio aggressively educated our employees, our customers and suppliers, and the communities in which we live and work.' The initiative was handily defeated as a result of Cargill's intervention.[46]

Cargill also used its grass-roots techniques to lobby for the North American Free Trade Agreement (NAFTA). After members of the Cargill Community Network had done their organizing work, employees at Cargill's 600 locations in the US were given information about the trade agreement and given cards to send to their congressmen. Cargill figures that well over 50,000 cards may have been sent to Washington. As Cargill told its employees: 'NAFTA is important to Cargill because it clears the way for what we do.'[47]

When William R. Pearce retired as vice-chairman of Cargill in 1993, the Minneapolis *Star Tribune* carried an unusually frank report on Pearce's career and provided a rare insight into how Cargill works. 'Perhaps he has had more influence on public policy than most elected officials, save presidents,' commented staff writers John Oslund and Tony Kennedy, 'but despite a lifetime spent influencing the private affairs of Cargill Inc., the domestic affairs of US agriculture, and the foreign affairs of the United States' most powerful friends and enemies, most Minnesotans have never heard of Cargill's retiring vice-chairman'.[48]

Pearce started work with Cargill, as one of its four lawyers, in 1952, moving to the public affairs department in 1957 and then becoming vice-president of public affairs in 1963. In 1971 Pearce was appointed deputy special representative for trade negotiations – which gave him the rank of ambassador – by President Nixon. Pearce took leave from Cargill to accept the job, according to Kennedy and Oslund. In this position, Pearce steered a trade bill through Congress that set the stage for US international trade policy for a generation. Kennedy and Oslund report former secretary of state George Schultz as saying of Pearce: 'He had an easy way of getting things done, and he got them done the way he wanted them done.' Cargill's own comment on this aspect of Pearce's work: 'As a member of the administration, Pearce shaped international trade policy.' Pearce rejoined Cargill in 1974.[49]

When the Soviet Union invaded Afghanistan in 1979–80, the Carter administration imposed an embargo on the sale of agricultural goods to the Soviet Union. Pearce argued that if the government insisted on an embargo, then it should buy the grain already on its way to the Soviet Union. The government agreed on the condition that it would cover only the companies' actual costs, but not their anticipated profit. President Carter also asked Cargill to halt all grain sales to Moscow through its foreign subsidiaries. Kennedy and Oslund reported that: 'after rigorous internal debate, led by Pearce, Cargill subsidiaries from Canada to Argentina honored the embargo'. Cargill's own telling of the story does not mention Pearce's role in obtaining the compensation, due or not. In a later issue of the *Cargill Bulletin*, the company simply reported that: 'By the fall of 1980 ... the government was in the process of assuming the contractual obligations for 13 million tons of corn and 4 million tons of wheat once destined for the Soviet Union.* Eventually, most of the corn contracts were sold back into the marketplace. The embargoed wheat, however, was purchased outright and the 4 million tons were placed in reserve.'[50]

Those with less of a financial stake in the affair had other comments:

> The USDA [United States Department of Agriculture] compensated Cargill and its colleagues for grain they had agreed to, but could no longer, ship. A 1981 report by the Agriculture Department Inspector General ... described possible manipulation by unnamed companies. Large amounts of grain were reclassified as bound for the Soviet Union and thus made eligible for compensation.[51]

A similar windfall scenario had occurred in 1971–2 when the Soviet Union made surprise and unprecedented purchases of massive amounts of US-subsidized grains. According to Richard Gilmore, the sale of wheat to the Soviet Union in 1972 cost the US $300 million in subsidies, most of which went to the largest private exporters. 'The windfall came from the fact that several firms, having made sales to their foreign affiliates before the government's notice of the forth-coming termination of the subsidy program, subsequently registered these sales at the peak subsidy rates.'[52] Cargill's sales (the figures include subsidies) went from $2 billion in 1971 to $29 billion in 1981. Figures are not available for the intervening years. One has to draw one's own conclusions.

The most high-profile Cargill executive to directly shape US policy as a member of the administration has been Daniel Amstutz. Unfortunately his name recurs all too often to the exclusion of others who have probably served Cargill's interests equally faithfully, not only in the US, but in many other countries.

Amstutz started his career with Cargill in 1954 as a grain merchant, moving up to the position of assistant vice-president for feed grains in 1967 and then on to the position of president of Cargill Investor Service in 1972 where he remained until 1978 when

* While distance and area are given in metric measurements throughout the book, weights have been left in their original form due to considerable inconsistency in industry reporting. Thus, weights are given variously, kilograms, pounds, tonnes, tons and metric tons – an incorrect term which has nevertheless been used by Cargill.

he left Cargill to become a partner in Goldman, Sachs and Company developing their commodities trading business. In 1983 Amstutz became US Under-Secretary of Agriculture for International Affairs and Commodity Programs and president of the Commodity Credit Corporation, all of which made him chief policy officer for US farm programmes. From 1987 to 1989 he held the rank of Ambassador as chief negotiator for Agriculture in the GATT negotiations. From 1989 to 1992 he was a private investor and consultant, at which time he was appointed executive director of the International Wheat Council. In 1998 he turned up as president and CEO of the North American Grain Export Grain Association.

Much of Cargill's lobbying – and that of all agribusinesses – is carried out through trade associations such as US Wheat Associates and the National Grain and Feed Association, and commodity organizations such as the National Corn Growers Association, the Canola Council and the American Soybean Association (it's not really an association of soybeans, but of corporate soybean interests such as big growers, processors, seed companies, and so on).

> The company prefers to work through influential trade associations [whose] state-of-the-art lobbying techniques offer Cargill the advantages of Washington influence without the costs of close company identification with controversial proposals or overt political tactics.[53]

While this was written in 1985, it is as true now as it was then, and the public can certainly be excused for confusing agribusiness lobbies with legitimate farmers' organizations. TNCs such as Cargill do nothing to ease the confusion. On the contrary, agribusiness corporations are highly skilled in representing their corporate interests as the interests of farmers. When agribusiness talks about agriculture, it is talking about a very specific form of capital intensive, industrial commodity production. Each commodity is described as an industry – the sugar industry, the cotton industry – and the categories are considered inclusive of everyone from grower to trader to processor to manufacturer. This results in a wonderful mystification of an unacknowledged power structure. The powerful industry/processor members, though few in number, easily dominate the interests of hundreds, or hundreds of thousands, of farmers. Any legitimacy these groups might have as farm organizations depends on acceptance of the premise that processors and growers share common interests. Of course they do, to some extent, but they also have dia-metrically opposed interests: farmers need the highest price possible for their crop, while the processors regard the farmers' production as an input they want to acquire for as low a price as possible.

Capitalist Subsidies

When the government of the US entered the grain business in a serious way in the immediate postwar years, first through UN Relief and Rehabilitation Agency programmes and then directly with the Marshall Plan, Cargill had already been in the business for 80 years. These programmes moved mountains of grain aid to Europe, with the result that US wheat and flour exports jumped from 48 million bushels in 1944 to 503 million in 1948. The grain majors, including Cargill of course, were the agents of these programmes on behalf of the government, and as such they did well storing and delivering grain on a cost-plus basis.

By the early 1950s, however, Europe was on its feet, determined to become self-sufficient in food production after the trauma of hunger and food insecurity during the war and immediate postwar years, and grain imports were replaced with domestic production. The dumping of US grain in Europe was no longer welcome foreign aid, but unwelcome competition and an obstacle to the European goal of self-sufficiency.

The ingenious response of the US and its grain lobby was the passage of Public Law (PL) 480 – the Agricultural Trade Development and Assistance Act, known as Food For Peace – in July, 1954, that set US grain exports on an upward path again. PL 480 'combined and extended the use of surplus agricultural products for the furtherance of foreign policy goals ... The funds could also be used to develop new markets for United States farm goods ... That it was a boon to the American grain traders goes without saying', wrote W.G. Broehl in his history of Cargill.[54]

As an agent of the government, Cargill has always been one of the prime beneficiaries of PL 480 financing. At the same time, as a private trader, Cargill has benefited handsomely as Food For Peace grain exports whetted the appetites of many new potential customers for subsequent commercial sales. In fact, the promise of eventual commercial purchases was often a specific precondition for the food aid in the first place. Food aid, particularly wheat, was utilized much like infant formula: to create a taste and a market for a company's products for a lifetime.

Between 1955 and 1965, Cargill's US grain exports increased 400 per cent, with sales rising from $800 million to $2 billion. By 1963 PL 480 had generated sales for Cargill and Continental of $1 billion each. (This was for storing and transportation, not for processing or manufacturing.) In addition to its increased sales under PL 480, Cargill benefited from the government's grain storage programme. Between 1958 and 1968 it received some $76 million for storing grain, often in leased publicly owned terminals or terminals built with public funds. Cargill's reputation in the trade for manipulating government programmes to its advantage is extensive. As a Dreyfus executive in Winnipeg put it, referring to Cargill: 'The big don't get that way by waiting around for something to happen.'

In 1964 US policy shifted from subsidizing the storage of grain to subsidizing grain exports only. Subsidies were paid to the grain companies so that they could discount the price and sell grain below both the domestic price and the prevailing world-market price. While the savings from reduced storage costs were expended in subsidizing exports, the government, or national budget, benefited by the increased foreign-exchange earnings.

Dan Morgan, in *Merchants of Grain*, pointed out that officially all this was called 'making American agricultural products more competitive abroad'. Even the conservative *Financial Times* of London was explicit about the dependency of the private companies on public subsidies: 'At the height of US "grain power" in the 1970s, companies like Cargill Ltd and Continental Grain Co. made fortunes out of US agricultural exports. Privately owned and secretive, they are the two largest members of a group of five companies that controls between 85 and 90 per cent of US grain exports ... Fierce advocates of a free market for agriculture, they have become overwhelmingly dependent on Government efforts to increase their sales.'[55]

With PL 480 still in place and in use, in 1985 the Congress of the United States passed the Export Enhancement Program (EEP) of the Food Security Act, putting in place the most notorious of the publicly funded corporate assistance programmes. Neither the EEP nor PL 480 have improved the lot of farmers themselves. Under the EEP, eligible countries are designated year by year by the Secretary of Agriculture. Individual sales are then negotiated between the eligible country or its designated agency and one of the trading companies on the basis of the subsidy available at the time for that particular country. The subsidy is then paid, in one form or another, to the company making the deal.

In its first four years of operation (1985–9) the EEP had 'targeted' 65 countries with twelve commodities, including flour. Clayton Yeutter, who at that time was US Trade Representative, explained on many occasions that such programmes were necessary to counter the subsidized exports of the European Community and to subsidize US farmers so they could compete on world markets.* Regardless of the rationalization, the effect of the EEP has been to pull down the 'world market price' and, consequently, often with great damage to their domestic agriculture, the prices received by farmers in the recipient countries for their grains.

Who really benefits? In 1987 it was reported that wheat sales to China under the new EEP netted Cargill bonuses worth $2 million, while Dreyfus and Continental each benefited by half that amount, and during these years of the pro-business free-enterprise Reagan regime, grain traders Cargill, Dreyfus, Continental and Artfer Inc. (owned by Ferruzzi Group) collected $1.38 billion from the US government, more than 60 per cent of the subsidies through the EEP in its first four years. In other words, while condemning the 'trade distorting practices' of, for example, the Canadian Wheat Board (CWB) or the Common Agricultural Policy of the European Union, the US became a heavily subsidized *de facto* state trading corporation.

To dispel any doubt as to the intent of the EEP, in 1989 the USDA issued new guidelines for the EEP, two of them making the purpose of the EEP quite explicit: EEP proposals must further the US negotiating strategy of countering competitors' subsidies and other unfair trade practices by displacing exports in targeted countries and all EEP initiatives must demonstrate their potential to develop, expand or maintain markets for US agricultural commodities.[56]

In its first year, the EEP accounted for only 12 per cent of the 25 million tonnes of US wheat exported, but by 1987–8 this had climbed to 70 per cent of the 45 million tonnes exported. During this same period the debt of the third world or 'less developed'

---

* The office of The US Trade Representative was created in the Trade Expansion Act of 1962 to represent the US in trade agreement negotiations and to administer the trade agreements programme. The Trade Act of 1974 expanded the Special Trade Representatives's responsibilities, gave the office Cabinet-level status and gave the trade representative the rank of ambassador.

countries was growing and they were increasingly unable to afford the grains that the US needed to sell. Since 'the market' was unable to play its part in moving US grain surpluses, government intervention was a growing necessity.

Three extensive articles in the *New York Times* in 1993 evaluated the EEP: 'The agriculture Department's $40 billion campaign to bolster crop exports, begun a decade ago to help beleaguered farmers, has instead enriched a small group of multinational corporations while doing little to expand the American share of the world's agricultural markets ... An examination of the subsidy programs highlights the symbiotic relationship between one of the biggest and least scrutinized federal departments and some of the politically influential companies it regulates.[57]

PL 480 and the EEP are not the only publicly funded programmes that have benefited the grain processors and merchants in the name of US market share and global competitiveness. Programmes such as the Targeted Export Assistance Program and the EEP are often channelled through industry foundations and associations. US Wheat Associates and the US Feed Grains Council, for example, are among 46 organizations that have received suport from the Targeted Export Assistance. One project undertaken by US Wheat Associates sent 100-tonne samples of various classes of US wheat to mills around the world along with US specialists who worked with the potential foreign users. Mills in Senegal, Burkino Faso, Colombia, Taiwan, and many other markets have participated in this programme. In 1988 more than 1,000 small bakeries in Korea participated and ten new baked foods were introduced.

In 1989, The National Association of Wheat Growers Foundation developed a project called The Developing World: Opportunities for US Agriculture with the intent of increasing opportunities for US wheat exports to less developed countries. 'The project will train up to 30 growers to make presentations to state and local groups, and through the media, on economic development and trade and the potential of less developed countries to enhance the US economy.'[58]

The USDA Market Promotion Program (MPP) provides funds or commodities owned by the Commodity Credit Corporation to trade organizations, companies and cooperatives to implement foreign market development programmes. US Wheat Associates has used MPP funding for a number of years for projects such as flour-milling schools in Egypt, Venezuela and Morocco and baking schools in Thailand, Costa Rica and Algeria. US Wheat Associates' president

Winston Wilson said: 'The US presence in the school will broaden familiarity with US wheat and maintain relations with Latin American milling industry representatives in the face of aggressive competition from Canada and other world suppliers.' Cargill, as the major player in the flour and pasta industry in Venezuela, has certainly been a primary beneficiary of this programme.

While Cargill was, typically, utilizing the good offices of the US government under the guise of the EEP and other programmes to intervene in the world grain market for the sake of sales and market share, it was also trying to destroy other state trading companies, such as the Australian Wheat Board (AWB). As a member of the Australian Grain Exporters' Association, a coalition that included ConAgra, Continental Grain and Louis Dreyfus, and with the encouragement of the Australian government, Cargill sought to break the export monopoly of the AWB by promising higher grain prices to the farmers and lower costs to the government if the trade was deregulated and thrown open to the private traders. It is exactly the same game Cargill plays in Canada against the CWB. In 1992 the AWB was told it would retain its export sales monopoly until June 1999 and then it would become a quasi-government organization with limited monopoly powers. The government said it would continue to guarantee 85 per cent of the board's borrowings to finance export sales of grain.

A long-term replacement for the withered market of the former Soviet Union (FSU) is obviously needed by the US and Canada. China is the only possible one, thanks to its sheer magnitude. For this market to be exploited, however, it must not be offended. But it must also have more adequate infrastructure to handle large grain shipments – just the sort of project that the World Bank favours. True to its purpose of financing agricultural 'modernization' and infrastructure for use by TNCs, in 1994 the World Bank announced a $1 billion (said by China to be $1.75 billion) investment programme, the largest such undertaking in history, for the construction and installation of modern grain-storage depots and handling equipment at 370 interior and export–import sites throughout the country. The largest port facility is to have 300,000 tonnes of grain storage capacity. One of the major aims of the project is to convert grain handling in China from sacks to a bulk system. Many of the new port facilities are to be designed for both import and export movements.[59] What more could Cargill ask for?

The north-east of China is reported to be extremely fertile and capable of producing large quantities of maize and soybeans. Port facilities are to be built on the north-east coast to facilitate the export of these crops and the import of wheat. Port facilities are also to be built at the mouth of the Yangtze river so that maize and soybeans can be imported from the north-east as well as wheat from around the world. Port, storage and distribution facilities will also be built in the south-east to serve the heavily populated 'southeastern corridor'. Once these facilities are in place, China will be thoroughly integrated into the global grain system of the 'grain majors', as the Japanese describe Cargill, Continental and others. Cargill, of course was one of the strongest advocates of bringing China into the World Trade Organization (WTO), which happened in 2001.

5   Creatures: Feeding and Processing

Livestock of one sort or another were once to be found on virtually every farm in North America, and, as long as the grass was still growing and the snow was not too deep, the livestock collected and processed their own feed, supplemented with the leftovers of human food processing and meals. Not so any more. Poultry and pigs are raised almost totally in confinement with all their feed brought to them, dairy is increasingly the same, and beef is a combination of fairly extensive cow–calf operations and totally confined feedlot growing and 'finishing'. It made eminently good sense then for Cargill to counsel this change and to position itself in the middle of this as a feed manufacturer as the transition occurred, first in North America and increasingly around the world. Where once the cows grazed, the poultry scratched, and grain was produced for human consumption, now grain and oilseed production have vastly increased, mostly for animal feed. Cargill has encouraged and exploited this transformation with huge success.

Cargill entered the formulated feed business in the late 1930s and got into slaughtering and processing cattle, pigs and poultry and the milling of corn and soy in the 1960s. When it did integrate its extended lines of business, from seed to feed to slaughter, Cargill was in a position to take full advantage of the resulting synergies and financial efficiencies. The process appears to be a combination of the rational pursuit of profit and growth combined with skilful analysis to identify both new lines of opportunity and dead ends.

Will Cargill had sold simple feeds – milled grains, essentially – as early as 1884, but a more serious start was made in 1934 in Montana with Cargill brand feeds. Five years later Cargill entered the formula feed business with the construction of new buildings in Lennox, South Dakota, specifically for the purpose of feed manufacturing. The product was marketed under the brand name Blue Square. This business expanded rapidly into Minnesota and Iowa and was run for some time quite separately from the Montana feed division.

At the end of World War II, Cargill made two big acquisitions in feed milling and oilseed processing: Honeymead Products Co. of

Cedar Rapids, Iowa, which included a feed plant and a soybean processing mill, and Nutrena Mills in Kansas.

Honeymead was owned by the Andreas family, and when it was bought by Cargill it was being run by a young member of the family, 27-year-old Dwayne. Dwayne Andreas stayed with Honeymead and Cargill, soon becoming a vice-president. He left Cargill in 1952 in a dispute over management styles and went to work with the Grain Terminal Association. In 1966 Dwayne Andreas and his younger brother Lowell were invited by the Archer and Daniels families to become majority shareholders of their company, ADM, and run it. In 1971 Dwayne became CEO of the company. Under the leadership of Andreas, ADM became a major competitor to Cargill in oilseed processing, flour milling, grain handling and other activities, though with a very different management style that was very aggressive and blatantly political. Andreas, now 83, was head of ADM, as chairman of the board, until 1999.

The purchase of Nutrena Mills, then a major mid-west milling company centred in Kansas and offering a complete line of poultry, dairy and pig feeds, cost Cargill $1.6 million and took the company directly into the retail world. Nutrena became Cargill's feed division and Cargill's Montana-based Blue Square brand was replaced by the Nutrena label.

The company's next expansion in the feed business came six years later with the acquisition of a major interest in Royal Feed and Milling Co. of Memphis in 1951. Cargill rolled it into Nutrena and in the years following continued its expansion in different regions of the country, acquiring Beacon Milling in 1985 and buying Acco from Quaker Oats in 1987.

Cargill extended its Nutrena feeds business into the US north-west with the purchase in 1989 of Hansen & Peterson in Washington state. This gave Cargill a total of 58 mills in Canada and the US. The Washington state acquisition complemented Cargill's big mill in Stockton, California, and suggests that Cargill believed that the dairy and cattle business would continue to concentrate in the west. Further evidence of this is provided by the company's construction of the biggest feed mill it had ever built, a $10 million mill at Hanford, California, in 1992 to produce dairy feed primarily. That same year Cargill began building a feed mill at Wooster, Ohio, and completed the purchase of a feed mill in Louisiana that serves the aquaculture industry as well as supplying bagged animal feeds.

Cargill continued its gradual expansion in the feed business, both in the US and around the world, and had 100 or so feed mills in the US and several other countries when it made a quantum leap in 2001 with the acquisition of Agribrands International Inc. for $535 million. Agribrands had previously been spun off by Ralston Purina Co. along with Ralcorp, which makes private-label breakfast cereals and crackers. Agribrands produces feed for everything from pigs and rabbits to shrimp under the Purina and Checkerboard names from 71 plants in 16 nations. Cargill feed is sold under brand names that include Nutrena and Acco. The acquisition of Agribrands, as Cargill notes in its corporate brochure, doubled Cargill's presence in the global 'animal nutrition' market, adding strength in aquaculture, the fastest growing segment of the feed industry. Cargill emphasizes tilapia and shrimp, both of which bear much more resemblance to factory poultry production than to traditional livestock tending.

Cargill LiquaLife is the world's first liquid shrimp feed. Each drop contains nutrient beads and direct-fed microbials suspended in a liquid medium – providing highly available nutrients to larval shrimp while helping prevent the accumulation of ammonia nitrogen in culture tank water.[60]

If you are in the feed business, you are already very close to the animal feeding business, so it is only logical that Cargill has long been involved more directly in animal feeding.

In 1980 there were 78,000 feedlots in the 13 major cattle-feeding states of the US. By 1992 this number had dropped to 46,450, although the number of cattle on feed remained relatively constant as the feedlots got bigger, and this trend continues, both in the US and Canada.

In 1996 the top ten cattle feeders in the US could handle 2.5 million head of cattle at one time in 50 separate feedlots. Continental Grain led the pack with space for 400,000 head at any one time and an annual total estimated at 975,000 head. Cactus Feeders was number two with 330,000 head capacity at any one time, ConAgra number three with a capacity of 300,000 and an annual total estimated at 800,000, and Caprock Industries, Cargill's appropriately named ('industries') cattle-feeding business, was number four with a one-time capacity of 285,000.

You want to go back to the 19th century? You want to have a packinghouse in every little town and deal with 21st century marketing? There's no way! ... There is no stopping it. This is an evolution that's going to take place in spite of whoever is in the way. – Robert Peterson, chairman and CEO of IBP Inc.[61]

The development of feedlots (invented in the 1950s as a mechanism to use up the surplus grains that were no longer welcome in Europe) and their concentration in the mid-west of the US and Alberta in Canada was accompanied by the relocation of the processing facilities (what used to be called 'slaughter houses') from population centres, large and small, to the centres of cattle feeding. Not long after this transition, another took place in the final journey of the deconstructed animal. Instead of shipping beef by rail as 'swinging' beef (split carcasses hanging from a hook) in refrigerated railcars from western slaughterhouse to eastern market (which providing time for the meat to 'age'), the carcass was broken down into large chunks at the slaughter house and packed in boxes for shipment by truck to the wholesalers and retailers who did the final cutting on site.

The latest development, which once again facilitates the concentration and centralization of the meat-packing industry, is the preparation of 'case-ready' retail cuts at the site of slaughter or in a subsidiary facility. 'Case-ready' means that the meat is fully cut, wrapped, packaged and, in some cases, even weighed and priced, by the processor/wholesaler.

Fortunately, feedlots and packing houses are not the totality of the beef industry. There are the cow–calf farms (the source of feedlot cattle) which utilize the natural capacity of ruminant animals to graze and feed themselves and there are farms that raise essentially grass-fed and organic beef. These types of enterprise can be found almost anywhere but they are almost all small-scale compared to the rest of the livestock 'industry'. Cargill Canada's long-time vice-president Dick Dawson, who retired in 1993, once said of Canada and the meat business:

More people all over the world are living better today. Our challenge is to grow more tons, value add to more meat, value add again to more further processing and selling into a rising market ... More people live at even higher standards the world over. More people also starve. All the graphs point upward. We are in an irreplaceable business on a growth trend.[62]

Cargill launched itself into beef slaughter and processing in a big way in 1968 with the purchase of MBPXL for $68 million. The company was renamed Excel. By 1991, Cargill owned 31 meat- and poultry-processing plants throughout the world. It operated 14 beef and pork plants and three broiler chicken and three turkey-processing plants in the US plus the largest beef-packing plant in Canada at High River, Alberta. By 1991 it was also operating a broiler-processing plant in Bangkok, Thailand; a beef and sheep plant in Australia; a pork-processing plant in Taiwan; a beef- and chicken-processing plant in Saultillo, Mexico; a beef plant and a poultry plant in Honduras; and a broiler plant in Argentina. Its Sun Valley subsidiary operates a broiler-processing plant in England and a turkey plant in Wales. In 1999 Cargill added one of Costa Rica's largest meat processors, Cinto Azul SA, to its stable of meat processors.

Lest Cargill be given credit for doing it all on its own, it is good to note the public assistance it has received for just one project in Nebraska:

> In 1993 Cargill decided to replace its existing meat processing plant in Nebraska City with a new $15 million facility. Unwilling to finance this small project on its own, it asked the Nebraska department of economic development for a $1.55 million grant, the highways department for $304,000 in road improvements, and the Federal Economic Development Administration for $445,000. It then asked the residents to vote themselves a $2.63 million tax hike to finance the plant.[63]

Cargill's entry into and domination of the Canadian beef sector provides a good portrait of the company's strategy and practices.

In 1989 Cargill opened its $55 million beef-packing plant in High River, Alberta, (with the help of $4 million from the Alberta Government), with wage rates in the plant about $2.50 below the rates in other western Canadian plants at the time, no union, and a kill rate of 1,600 per day five days a week, single shift.

By the time Cargill opened its plant, 700 workers at four Canada Packers' plants in Alberta had accepted a $1.50 per hour rollback in their basic wage to $12.51, the average of wages at other Alberta plants, including Cargill's. Canada Packers had forced this concession from the workers in 1988 under threat of closing the plants altogether. In other words, Cargill had effectively set the basic wage rates for the packinghouse workers in the province of Alberta a full year before it was actually in business.

A year after the plant opened the 430 workers in the bargaining unit at the High River plant voted for union certification. The agreement with Cargill called for a starting wage of $8 an hour, rising to $9.60 after one year and up to $10.95 as the top wage for a skilled worker. By mid-1993 Cargill's base wage rate had risen to $10.25.

When the High River plant opened, Lakeside Packers, a unit of Lakeside Farm Industries Ltd, of Brooks, Alberta, was Cargill's only potential competitor. Then IBP, the world's largest fresh-meat processor at the time, purchased Lakeside in 1995. Cargill's response was a $37 million expansion enabling the plant to run a full second shift, bringing the daily kill rate to 4,000 head.

In mid-2001 Tyson Foods bought IBP, including the Lakeside plant. This made Tyson the largest meat company in the US with sales of $24 billion and control of 28 per cent of the US beef market, 25 per cent of the chicken market and 19 per cent of the pork market. Cargill remains in third position behind ConAgra with sales for its meat division, Excel Foods, of about $10 billion, which represents about a quarter of Cargill's total business.

Once its plant in High River was in operation, Cargill wasted no time finding ways to get its product distributed in the lucrative Ontario market, including the use of the distribution system of Maple Lodge Farms, until it was able to purchase the Trillium Meats Ltd plant in Toronto from Steinberg, Inc (Soconav) of Montreal in 1993. Trillium was a meat-cutting and distribution operation that had been used to supply Steinberg's Miracle Food Mart stores before Steinberg's financial desperation forced the sale of its stores to A&P. Maintaining a corporate tradition, Cargill closed the deal with Steinberg only after the union members agreed to a substantial wage reduction rather than lose their jobs, so Cargill acquired both a cheap plant and cheap labour. Cargill viewed the plant as an experiment and an extension of its High River operation, with sides of beef and boxed beef trucked from High River to the Trillium facility for further processing into case-ready retail cuts. This has obviously worked well for Cargill. Early in 2002 it announced it would build a new $45 million case-ready meat and poultry plant in Chambley, Québec, with the help of 'a financial contribution (non-repayable)' of $3.6 million from Investissement Québec, $300,000 from Emploi Québec for workers' training, and the Municipality of Chambley will spend $350,000 to put in place the necessary infrastructure. Cargill has built a similar facility in Georgia to supply US supermarkets with case-ready meats.

In 2001 Cargill purchased Emmpack Foods of Milwaukee. Emmpack, a producer of 'value-added meat products', will enable Cargill/Excel to produce as much as 180 million pounds of cooked meats annually. This is really an extension of the case-ready concept into the home kitchen. Excel also converted its pig plant in Marshall, Missouri, to produce case-ready product and purchased Taylor Packing Co., a beef-processing facility in Wyalusing, Pennsylvania, which Excel planned to convert to a case-ready facility. The president of Taylor Packing Co., Ken Taylor, put a brave face on an all-too-familiar pattern of the little fish being swallowed by the big: 'The decision to relinquish control of a multi-generation family business has been extremely difficult. But the future offers exciting growth opportunities in our industry, and we are very pleased to meet that future as part of Excel ... For our suppliers, our customers, our employees and our communities, there is no company we would rather join.'[64] It comes down to a stark choice of 'sell out or be pushed out'.

When Cargill announced it would build the plant at High River, it was already the major feed and feed-supplement manufacturer and supplier in southern Alberta, and if a feedlot operator was short of cash, he could get financing to buy cattle through Cargill's Financial Services Division. The stipulation was that the cattle that were financed by Cargill had to be raised on Nutrena feeds. There really was no choice since no bank is eager to loan money for feed when the cattle are already assigned to Cargill as collateral for the loan with which they were purchased. If Cargill is then also the cattle buyer, it can tell the grower what specifications the cattle have to meet, when they will be shipped, and how much will be paid for them. The outcome is that the macho cowboy/rancher or feedlot operator is, in reality, little more than a franchise operator.

Altogether Cargill had an immense impact on Canada's beef production, effectively draining cattle off diversified farms throughout the west and into feedlots clustered in southern Alberta close to Cargill's High River plant.

XL Foods of Calgary, a relatively small beef processor in Calgary, Alberta, was one of the victims of Cargill's determinative presence. It placed the blame for its financial woes on Cargill, saying: 'The marketing strategy employed by Cargill, when matched with other packers attempting to retain market share, has totally destroyed margins, resulting in severe losses to the entire Canadian industry.'[65] XL sought to salvage its own finances by restructuring wages by means of a lock-out. Its workers were allowed to return to work only after accepting an average roll-back in wages of $2.39 per hour and a shorter working week. The company justified the cuts on the ground that it had to be competitive with Cargill.

Cargill also put tremendous pressure on the other packers in 1989–90 by paying top price for cattle in order to fill its plant capacity. In the short term, bidding-up cattle prices doesn't seriously hurt a company with Cargill's resources, but it may help drive its competitors out of business, particularly if the current wholesale price of beef is undercut at the same time. Buyers as far east as Nova Scotia reported that Cargill was 'low-balling' the wholesale market in late 1989 in order to get established. Cargill simply paid more for cattle and sold for less until it had the market share it wanted.

When Cargill opened its High River plant, Canada Packers was still the largest beef processor in Canada with three plants in western Canada killing 12,000 cattle per week. Then in mid-1990 Hillsdown Holdings plc purchased Canada Packers for a reported $700 million and the next year shut down two of its three beef plants. The third was sold to Burns Foods Ltd. Burns Foods did not fare well, and it too blamed Cargill for its troubles. Cargill, said Burns president Arthur Burns: 'is offering carcass beef at 10 to 14 cents a pound below the market, when normal profits are only 3 to 4 cents a lb'.[66] Cargill vice-president Bill Buckner responded: 'It's not true. North America is a very competitive marketplace. We've been pricing to make sure we can compete in it.'[67]

By no coincidence, the Calgary-based Canada Beef Export Federation was incorporated in 1989 as a non-profit federation representing beef packers, processors, exporters and provincial and national beef associations. The Federation engages in market research and sponsors trade missions. The organization is supported by contributions from provincial cattlemen's associations and a variety of public sources, both provincial and federal. Industry contributions, if any, are unpublished, but as the major exporter, Cargill is clearly the major beneficiary of the organization's activities, which parallel those of the US Beef Export Federation.

From Cargill's standpoint, it's a great system. Being the feed supplier, the banker, the buyer of the finished cattle, their butcher and their wholesaler creates a tidy system that gives the company maximum control and return with the major risks – weather and animal health – being shouldered by others. It is also a very good way to market cheap grain, as long as you are not the farmer who grows it.

The scale of contemporary North American beef production is as hard to imagine as it is to see. Getting a tour of a large packing plant to get a first-hand view of the deconstruction of a large animal is a virtual impossibility for reasons of health and safety and offensiveness. Beef feedlots are visible and visitable, but located only in very particular geographies beyond the travelling reach of most people. The only more or less ubiquitous and visible aspect of beef production in the mid-west is the fields of corn and grains that constitute the bulk of feedlot cattle feed. (The feedlot conversion ration is 8–9 pounds of grain to produce one pound of beef.) There is no better place to get a view of this than in central Nebraska. The feedlots – miles of them – are situated on a south-facing slope that follows the wide arc of the Elkhorn River in the northeast quadrant of the state. The flat bottom land of what was once a much bigger river is now corn, while the feedlots utilize what was once the river bank. The uplands are corn again. There is not much 'wasted' space, unless, of course, you count that taken up by the feedlots themselves. It is all a stunning example of monoculture syndrome.

The Cargill feed and feed-supplement business in the region is carried on both under the Cargill/Nutrena and Walnut Grove names. Walnut Grove was a local feed company for many years before it was bought by W.R. Grace. The three warehouse locations and one feed plant give no visual or print clues to Cargill ownership, having been purchased by Cargill from W.R. Grace in 1991. (In the same deal, Cargill also acquired Farr Better Feeds in Colorado from W.R. Grace.) I stopped by the Walnut Grove office and was a little surprised at how quickly the men working there expressed their unhappiness about working for Cargill. They thought working conditions were bad under W.R. Grace, compared to what they were when Walnut Grove was really a local company, but they were even worse under Cargill. They said that Cargill is arrogant and that salesmen have to have a university degree to get hired, but you can't sell feed just because you have a laptop computer, which is now mandatory. The men said they are not supposed to talk baseball or family with customers, just 'business'. But that doesn't sell feed – which is why they try to keep invisible their Cargill ownership. They told me that the company caps – the kind of baseball caps with corporate logos that company salesmen give away to their customers and that every farmer has at

least six of – used to cost the Walnut Grove salesmen $1. Cargill raised
the price to $5 and then gave the salesmen a 4 per cent raise, which
just about covered the extra cost of the caps. They also told me that
business went down when Cargill bought Walnut Grove because the
local people don't like doing business with Cargill. Cargill is just too
big, has too much power, they said. When Cargill bought the
business, they told me, it fired more than half the management. No
one can understand the accounting systems they are now forced to
use. I later discovered that Cargill feed dealers around the world are
supposed to be using the same system. There is resistance, however,
in very disparate places. I even heard a similar story in Taiwan. No
one at the retail level may understand it, and it may not sell grain, but
the uniform accounting language insisted upon by head office has
another purpose: 'One of the challenges in running a worldwide
business is making sure employees are talking the same language
when it comes to reporting on business performance.'

In 1990 Nutrena standardized its reporting forms throughout
North America, but separate computer systems were being utilized, in
the local language, in other 'geographies'. 'Not only was it impossible
for these locations to do all their business in English, but cultural differences and a large variety of local business practices made a
common system seem impossible.' By 1993, however, Cargill could
report that 'in the fishing village of Shibushi, on the southernmost
island of Japan' and in Nakom Pathom, Thailand, local employees
were working in their own languages but the figures they sent to
Cargill headquarters in the US were all in the same format thanks to
special computers that translate the information and store it in a file
from which it can be retrieved in any language, providing Cargill
with 'a flexible system that is the same throughout the world'.[68]

Months after my Nebraska visit, I heard that Cargill peremptorily
fired the Walnut Grove salesmen, with a pittance of severance pay,
yet, despite this, none of them were willing to send me the old copies
of the company magazines that they said they had at home.

Poultry

Virtually all poultry in the US is grown under contract to one of a
handful of processors. The Integrator, as these companies are called,
provides the grower with day-old chicks, usually from a company
hatchery, and supplies the feed, the medications, and the specifications of the required buildings. When they reach market weight the

Integrator buys back the 'finished' birds at a price and under conditions established by it. The growers provide the building and the labour and assume all the risks of disease and death.

Cargill entered the poultry-processing business with the purchase of a processing plant in Ozark, Arkansas, in 1967. In 1995 it spent $25 million expanding one of its plants in Georgia and announced plans to build a new $38 million broiler production and processing complex in Vienna, Georgia. It spent $25 million on the plant and then sold its entire US broiler operation, including attendant feed mills and hatcheries, to Tyson Foods in exchange for an undisclosed amount of money and a Missouri pig plant. The acquisition of Tyson's pig plant increased Cargill's pig slaughter capacity at the time by 50 per cent to 37,600 pigs per day, making it the fourth largest pork processor in the US and the fifth largest pig producer, with 77,000 sows in production.

When it quit, Cargill's broiler operations included four processing plants in Georgia and one in Florida and the company ranked about number 21 among poultry processors. 'We couldn't realistically expect to become an industry leader in the foreseeable future', said Cargill spokesman Mark Klein.'[69]

Turkeys, however, are a different matter. In 1998, with the purchase of Plantation Foods in Texas, Cargill, already the fourth largest turkey producer in the US, increased its turkey-processing business from 29 million to 37 million capacity. Cargill expanded its turkey-processing business again in 2001 through a three-way deal with Prestage Farms of North Carolina and Rocco Enterprises. Without taking into consideration the sales of turkeys originating with Prestage, the combined sales of Cargill and Rocco's turkey business will approach $1 billion.

The exploitation of both chickens and chicken farmers, characteristic of the 'modern' broiler industry ('industry' is indeed the appropriate term) has generated its own opposition in recent years as the power of the Integrators to silence the growers' complaints has finally been broken thanks to skilful organizing of the growers. For example, in March, 1992, Cargill settled with the US Justice Department, agreeing to continue contracting with Arthur Gaskins, president of the National Contract Poultry Growers Association, after cancelling his contract in 1989 when he and 30 other growers sued Cargill for allegedly under-weighing their birds over a period of eight years. Cargill had to agree that: 'it may not terminate the contract of any poultry grower because they participate in grower association activities, seek legal redress against Cargill, contact state or federal regulatory agencies or retain an attorney to represent them in any matter'. Previously, Cargill had held that it could terminate growers for any reason – or for no reason.

Cargill tried poultry in Canada, but in 1981 it shut down its Panco Poultry division in Surrey, British Columbia, after three years of operation. From 1965 until 1988, however, Cargill had major or total control of Shaver Poultry, Cambridge, Ontario. (Shaver's hybrid poultry, according to the company, are responsible for one-third of all white eggs produced in the world.) In 1988 Cargill sold Shaver to l'Institut de Sélection Animale (Mérieux Group) of France that already had the biggest share of the world's brown egg market. According to the press release, Cargill had concluded that poultry breeding is out of the mainstream of its integrated poultry operations. 'Cargill will concentrate future resources on live production, processing and marketing of poultry products.'

It did so 14 years later with the purchase of the chicken-processing plant in London, Ontario, and the hatchery operations in Jarvis, Ontario, of Cuddy International Corporation at the end of 2001. The terms of the transaction were not disclosed. 'We are extremely interested in Cuddy's chicken processing and hatchery business because of the opportunity it presents to strengthen our operations in the Canadian value-added food industry, as well as globally', said Cargill Limited (Canada) President, Kerry Hawkins. Cuddy is one of the world's largest supplier of turkey eggs and day-old poults, operating in Canada, the US and Europe.

Global Chicken

> Jas Matharu, a technical coordinator for Sun Valley, Cargill's European poultry processing business, works with more than 900 McDonald's restaurants in the United Kingdom that serve chicken nuggets and sandwich patties made by Sun Valley ... Matharu spends much of his time in McDonald's stores ... Besides inspecting all of the Sun Valley chicken products in the restaurant, he'll conduct a 30-minute food safety audit that includes measuring cold-storage temperatures. Then he'll go over cooking procedures with the restaurant staff and make sure cooked product isn't kept in warming trays beyond the McDonald's standard of 10 minutes. With Sun Valley's new precooking process for nuggets and sandwich patties, safety

concerns about undercooked chicken have been virtually eliminated ... With all the time he spends at McDonald's facilities, it can be difficult to tell whether Matharu works for Sun Valley or McDonald's. That's the way McDonald's likes it.[70]

While expressing its support for genetically engineered foods, Cargill does not hesitate to accommodate itself to market demands, such as excluding Roundup Ready transgenic soy from the poultry feed its Sun Valley subsidiary feeds McDonald's chickens. The non-GE soy has to come from Brazil, but the supplier is still Cargill. Walking on both sides of the road at the same time is a familiar practice for Cargill – as long as there is money to be made on both sides.

Cargill will handle whatever sells, such as organic and conventional grains, or genetically engineered crops and non, both in IP programmes, such as its InnovaSure programme. In 1999, then chairman Ernest Micek told a business breakfast meeting: 'People want the freedom of choice', and therefore, 'we need to go through this system of labelling and identifying GMO and non-GMO materials'.[71]

Given the highly concentrated character of the US poultry business and the near-saturation of the market, it is hardly surprising that Cargill looked elsewhere for more rapid expansion in one of its 'core competencies'. As early as 1970 Cargill was attracted to Indonesia by the country's large population and agricultural economy and sent one of its experts to study the situation. The recommendation of Cargill scout Kees Nieuwenhuyzen was that Cargill start a feed company and a small breeding hatchery, and build from there. 'For a company of Cargill's size, the start was very circumscribed: a small labour-intensive factory 60 km outside Jakarta that cost $250,000 with a capacity of 200–300 tons of feed per month.' At the time Cargill owned Shaver Poultry in Canada which could supply the breeding stock for the poultry operation, thus enabling Cargill to supply farmers with the chicks and their feed.[72]

By 1982 the operation had grown to two feed mills, three chicken-breeding farms and a hatchery with an annual production of 4.5 million broiler and layer chicks. Hybrid corn seed, which had been developed by Cargill in and for Thailand, had also been added to the company's products. The seed, it was said, worked so well in Indonesia that the government decided to subsidize 30 per cent of the cost of the seed to farmers. 'We didn't ask for it, we wouldn't have asked for it ... but we can't say no. So we make the best of it.

And the important thing is not the subsidy itself but that the government indirectly becomes a vehicle for us to get the seed sold', said Nieuwenhuyzen.[73]

From there, Cargill opened an office to allow better contact with the regional office in Singapore which had been opened in 1983 'and ease Cargill into a stronger position as an exporter of Indonesian products, primarily copra (coconut), tapioca, rice bran, and other grain substitutes in livestock feed, to Taiwan and Korea'.[74]

At the time, James Spicola was president of Cargill and he described the strategy Cargill applied in Indonesia:

> It's similar to the development we've followed in other countries. We start out with a reasonably small capital investment in a field to which we think we can bring some expertise and technology and management, then grow the business from there. We reinvest the profits and move into other opportunities as the situation develops ... We've found that our welcome to the country is much more productive on a long-term basis if we've started small and grown.[75]

Having established its welcome, Cargill embarked on a much more ambitious project in Indonesia: the development of integrated palm-oil production on Sumatra Island from the ground up under the name Hindoli. The crushing mill is to be one of the largest in the world. As the company reported it on its website:

> In the scrub land of southern Sumatra Island, Cargill is carving out its first palm plantation and building a plant that crushes palm fruit for its oil, which is widely used for cooking throughout Asia. The $45 million project – Cargill's largest in Asia to date – will take six years to complete and involves planting more than a million palm trees. The mill will crush oil from these trees, as well as from another 2.4 million trees grown on land owned by families who live near the mill site.[76]

It has not been possible to obtain further information about this project since it was reported in 1997. However, the general situation was described at the time in news reports.

> Because of a chronic urban crowding problem, Indonesia's government has systematically moved people from cities on Java to less populated islands. Since 1970, this voluntary 'transmigra-

tion' programme has resettled more than 7 million people into 2,600 villages, including 8,500 families who are part of the palm oil project.[77]

In September, 1997, Indonesia was blanketed by smoke from hundreds of thousands of hectares of burning scrub and forests, centred on western Borneo and eastern Sumatra. Indonesia's director-general for forest protection said the fires were burning on 164,000 hectares of land, of which 79,000 hectares were being cleared for plantations, 15,000 hectares were being worked by timber concessionaires and 70,000 hectares were natural forest. The central government wanted a huge expansion of plantation output in timber products, palm oil and rubber. To meet these goals, state-run and private companies – many with close business ties to the Suharto regime – were allocated vast tracts of public land for almost nothing and, as in every dry season, workers were burning those new lands to clear them for planting.[78]

Two years later, *Milling & Baking News* reported: 'Cargill has taken the initiative on alleviating some urban crowding and food security problems by investing in a palm plantation in Sungai Lilin, which is located approximately 120 km north of Palembang, Indonesia. In addition to the investment, Cargill is building a palm oil plant that will provide jobs to 8,500 persons ... Cargill's project is typical of the type of program that the United States hopes to see more of in the coming years.'[79]

In Thailand, Cargill set off in a different direction. Rather than starting very small and growing, it formed a joint venture in 1989 with Nippon Meat Packers of Japan to produce, process and market fresh-frozen chicken as Sun Valley Thailand, with facilities located in Lopburi and Saraburi provinces. The fully integrated business has its own parent-stock chicken farm, hatchery, broiler chicken farms, feed plant and broiler-processing plant. Operations began in 1990 with the company putting out something like 175,000 chicks weekly to the farms that grow them out. No further information is available.

Sun Valley Thailand's parent is Sun Valley Poultry Ltd of Britain, which Cargill acquired in 1980. Sun Valley sells one out of every four 'further-processed poultry products' eaten in the UK and provides McDonald's with all of the chicken nuggets and sandwich patties sold in the UK and most of western Europe. It also marketed chicken under such well-known private labels as Sainsbury's and Marks & Spencer. Sun Valley Foods employs 3,700 people in its chicken- and turkey-processing plants in Wales and in Hereford, England, producing poultry products for both the domestic and export markets. Altogether, Cargill employs almost 10,000 people in its international poultry operations located in five countries on four continents.

Eggs

Hens were once able to fend for themselves pretty well (except for the foxes), feeding in the barnyard and garden, pecking and scratching for fresh grubs and table leftovers. The farmer's job was to collect the eggs. By 1970 that quaint, though efficient, model had been replaced with capital-intensive industrial production and one worker tending 10,000 laying hens. Twenty years later that number was up to 100,000, thanks to automatic feeders and egg-gathering equipment. Cargill, which had been aggressively expanding in the fresh egg business in the previous decade, was described in 1987 as the largest egg producer in the US with 12 million layers under contract.

Like other integrators, Cargill contracted with farmers to grow Cargill-supplied chicks and then contracted with another set of farmers to look after them as layers. The feed, which accounts for about 60 per cent of an egg's production cost, was supplied, as part of the deal, by Cargill's Nutrena division, itself one of the country's five largest feed companies. 'Cargill literally makes up on volume what it loses on every chick because it makes a profit producing the egg' (referring to the money it makes on the feed it sells to the farmers) reported *Forbes*, which figured Cargill had 4 per cent of the US egg market at the time.

In 1989, Cargill sold the egg-production business of its Sunny Fresh Foods to Cal-Maine Foods of Jackson, Mississippi, making Cal-Maine the biggest egg producer in the US.

According to the feed industry trade magazine *Feedstuffs*, Cargill sold its shell-egg operations because shell eggs no longer fit its long-term strategy. The president of Cargill's Worldwide Poultry Operations explained that the company had been repositioning its poultry operations away from commodity products to further-processed, value-added products, such as liquid pasteurized eggs and cooked-egg products that are sold to food manufacturers and institutions. These products, which bear a striking similarlity to many other 'invisible' products that Cargill manufactures for the food industry, continue to be produced by Cargill's Sunny Fresh Foods, an egg-processing business in Minnesota which Cargill bought in 1985.

Since then, Sunny Fresh has added facilities in Iowa, Michigan and most recently, Ontario, Canada. The Canadian operation is a joint venture with the Kwinter family named EggSolutions Inc. Sunny Fresh now has more than 20 per cent of the egg further-processing business in the US producing more than 160 egg products primarily for the foodservice industry. These range from hard-boiled eggs to more than 30 versions of omelettes and more than 25 types of French toast. The biggest category is liquid pasteurized eggs, both fresh and frozen. About a third of the eggs consumed in the US are further processed, and Sunny Fresh goes through 2.5 billion eggs a year. Among its customers are McDonald's, Pizza Hut and Burger King as well as schools.

Catfish

Chickens and catfish don't appear to have much in common, but from the perspective of a feed company there's not much difference except one's in a cage and the other is in a pond.

Cargill entered the seafood business in 1989 when the Fishery Products Department was created to buy farm-raised shrimp overseas and sell them to food-service customers in the US. Two years later Cargill entered the US domestic fish industry when it began operating a leased catfish-processing plant at Wisner, in northern Louisiana. In order to get more control over the input side of the business, in 1992 it added 480 hectares of catfish ponds in southern Louisiana bayou country to the operation.

In its employee magazine, Cargill described how it worked with about 100 independent catfish growers while providing a buffer supply in its own production ponds at Lebeau, where there were 100 ponds each covering about 4 hectares. The fish were fed twice a day with puff pellets of feed mechanically sprayed onto the ponds' surface. As many as 50,000 to 70,000 pounds of fish could be caught in a single net. They were then trucked live, in a tank, to the processing plant at Wisner. Immediately after leasing the Wisner plant in 1991 Cargill began a $2 million expansion of its automated processing capability to get the plant up to the sanitary level that would give fresh product a guaranteed shelf-life of ten days. The Louisiana Commissioner of Agriculture announced that he was pleased and that this was 'an indication of a real commitment on the part of Cargill'. Two years later, a company called SF Services announced that it was going to purchase Cargill's catfish-processing plant in Wisner for $3.2 million. It wasn't really a purchase, however, because Cargill wasn't really the owner. Cargill had acquired the plant in the first place on a lease–purchase agreement with the Louisiana government, agreeing to pay $2.16 million for the plant over a ten-year period. SF obtained a lease–purchase agreement with the Louisiana government similar to the one Cargill had given up.

Cargill had apparently not attached enough importance to the existing relationships between processors and suppliers or fish growers. According to the trade magazine *Meat & Poultry*, the catfish industry in the US is unique for its concentration and organization, with four companies processing 90 per cent of the catfish. In addition, many farmers have a stake in the company which processes and markets their fish.

The truth of the matter came out in December, 1995. Faced with a jury trial in federal court in January, ConAgra, Hormel and Delta Pride Catfish Inc. agreed to a $21 million out-of-court settlement in a nationwide catfish price-fixing case without admitting guilt. ConAgra agreed to pay $13.6 million and Hormel $7.5 million, while the amount agreed to by Delta Pride was not revealed. 'The alleged conspiracy involved seven companies and went on for nearly a decade, during which time catfish wholesale prices were often remarkably similar throughout the industry, plaintiffs alleged.' Four of the smaller defendants had previously settled.[80]

Apparently Cargill was neither invited to join the party nor able to crash it. The conspiracy came to light in 1992 when the government initiated a grand jury investigation, just while Cargill was trying to get established in the business. Was it Cargill that blew the whistle? Or did Cargill just take its marbles and go home, saying that it would take more money than Cargill was willing to invest to play the game. As Cargill's Mark Klein put it: 'We got into the business as an experiment to determine if opportunity existed for us. We found we would have had to invest significantly more capital into the project and we couldn't justify it.'[81]

6 Cotton, Peanuts and Malting

Cotton

Although neither a grower nor processor, Cargill is nevertheless a major presence in world cotton trading through its subsidiaries Hohenberg and Ralli Bros & Coney. Cargill's involvement in cotton goes back at least to 1910 when the company established relations with Cotton Ginner Ltd in what is now Malawi. Someone within Cargill certainly knows the full story, but in his nearly 900-page history of Cargill, W.G. Broehl says nothing about Cargill's involvement in cotton and neither of its cotton-trading subsidiaries are mentioned.

Given that typically about 40 per cent of the US cotton crop is exported, with a value of $2.5 billion, and accounts for some 30 per cent of the total world export trade in cotton, it is hardly surprising to find four of the five major global cotton traders based in the US:

- Allenberg Cotton Co. of Cordova, Tennessee, a subsidiary of Dreyfus
- Dunavant Enterprises Inc. of Memphis, Tennessee, a family-owned company
- ContiCotton of Fresno, California, a subsidiary of Continental Grain Company
- Ralli Brothers & Coney, a division of Cargill PLC (UK)
- Hohenberg Bros Co. of Memphis, Tennessee, a Cargill subsidiary.

Hohenberg trades American cotton on both domestic and foreign markets from its Memphis base and operates offices in places like El Salvador, Guatemala and Mexico. Ralli Bros & Coney trades non-US origin raw cotton worldwide from its UK base. 'Hohenberg manages risk and manages supply for customers who want just-in-time delivery of very specific qualities', said Hohenberg manager Craig Clemmensen in Memphis.[82]

The US cotton harvest averages 14.5 million bales or 6.7 billion pounds of cotton (one bale weighs 500 pounds, or 226.5 kg) from some 4.8 million hectares of land. US cotton mills utilize some 4 billion pounds of cotton fibre annually. In addition to the cotton fibre, the cotton harvest yields whole cottonseed and cottonseed meal which are used in livestock feeds and cottonseed oil which is used in food products. Being a manufacturer of livestock and poultry feed, Cargill can take advantage of its trading activities to supply its feed business with cottonseed cake and oil.

An unexpected treat while visiting the Hohenberg office in Memphis was a look at their cotton grading 'station'. This is essentially a large room with very special lighting that faithfully reproduces the conditions of natural daylight. Every bale of cotton that office buys and sells is sampled and the fibre graded for staple, colour, size and overall quality. Every bale is identified as to grower and growing location. Buyers are thus able to specify exactly what kind of cotton they want and Cargill can deliver according to very precise specifications. The grading is still done by hand by highly skilled graders, hence the need for 'natural' light. Only the tensile strength is electronically tested. In the midst of globalization, electronic communications and corporate oligopolies, this reliance on human skill to maintain a quality basis of trade seems strangely old-fashioned, yet somehow comforting. Trading coffee on the basis of actual sample tasting is the only other example I know of such trading in real commodities, though Cargill has developed, as already described, its InnovaSure system for Identity Preservation of crops.

Yet Cargill was also pleased that its cotton trading subsidiary Hohenberg had overcome tradition and convinced some buyers to purchase large amounts of cotton on an 'undifferentiated basis' within a given range, rather than by specifications bale by bale as described above. This is like buying grain in the US on the basis of an average quality, rather than from the CWB on the basis of a specific uniform grade. Of course it is to the trader's advantage, not that of either the buyer or the seller, to work on an 'average' basis. Floor sweepings, gravel and dirt in general can gain commodity status if what they contaminate, or dilute, is of high enough quality to average it out.

The former Soviet Union was a major cotton producer and its collapse brought out the vultures. Since Uzbehkistan produced 65 per cent, or 1.6 million tonnes, of the total Soviet cotton crop, it was perfectly logical for Cargill's Ralli Bros division to open a cotton trading office in Tashkent, Uzbehkistan, in late 1991. 'We have to remember that the Soviet Union has been, over the past 27 years, the most important single trading partner of Cargill', Cargill International chairman Leonard Alderson told Cargill employees at the end of 1990.

Talking about Cargill's strategy for the FSU, Cargill vice-president Dan Huber said: 'We're going to focus our efforts on what we know ... We will start with projects in our core competencies – projects like seed, oilseed crushing, fertilizer production, application and distribution, feed manufacturing and grain warehousing ... We will also focus on the key agricultural areas or zones ... We will insist on having management control of any ventures.'[83]

In mid-1993 Cargill formed a joint-venture company called Den in the former Soviet Republic of Kazakhstan. The plan was for Cargill to hold 47 per cent of the company while the Kazakh trade company would hold the majority share. While this is contrary to Cargill's policy of holding majority interest in its joint ventures, in this case it means that Cargill 'has an early toehold in one of the most agriculturally rich CIS [Commonwealth of Independent States] republics'. The Kazakh government approved a resolution that allows Den to buy surplus commodities – left over after all state orders have been filled – from the republic's farmers. The government also ordered the state grain-purchasing agency to turn over to Den grain elevators having a total storage capacity of 600,000 tonnes.[84] Cargill partially explained the reasoning behind its commitment to such an enterprise in its *Bulletin*: 'Kazahkstan is attracting foreign investment ... partly because the old centralized government structure remained in place.'[85]

Cargill has always been able to admit to some of its mistakes and make a tactical retreat when conditions seemed appropriate, as with catfish, fresh fruit and the Japanese beef business. So in spite of long-time involvement, in 1993 Cargill decided that its Africa Division ought to sell its 40 per cent interest in the largest cotton-ginning business in Nigeria, Cotton and Agricultural Processors. The company's market share had dropped from 80 per cent to 47 per cent in the previous three years and its poor financial condition had been aggravated by a large loan from the state that Cargill headquarters did not approve of. The managing director of Cargill Ventures Nigeria said there were lessons to be learned, including: 'Stay away from minority businesses which Cargill does not manage; do not burn any bridges when divesting; and, be selective when choosing partners and try to avoid parastatals, which are bureaucratic.'[86]

Besides which, there are other cotton interests to pursue, like China. Cargill ships cotton by the container load to China from

Galveston, Texas, among other places, and according to one of its managers, handles 25 per cent of China's cotton trade.[87]

Among its cotton-related holdings Cargill had a cottonseed company, which it sold in 1994 to Delta and Pine Land Company of Scott, Mississippi, in order to be able to focus its efforts on corn, sorghum, sunflowers, canola and alfalfa. Cargill may have chosen to abandon cottonseed breeding to Calgene Inc., which owned Stoneville Pedigree Seed Company in Mississippi, and take on crushing seed for Calgene instead. Cargill subsidiary Stevens Industries, which processes peanuts, had already agreed in 1992 to process specialty canola oils for Calgene.

It was not long before Cargill exited the seed business altogether. In 1998 it sold its international seeds business to Monsanto, which had, several years earlier, acquired Calgene and had been trying to buy Delta and Pine. Then Cargill sold its North American seeds business to Dow in 2000.

Cargill has had operations in Africa for many years, with facilities in Egypt, Ethiopia, Kenya, Malawi, Morocco, Nigeria, South Africa, Tanzania, Uganda and Zimbabwe. An insight into Cargill operations there was provided by Cargill CEO Ernie Micek in a presentation to a US Senate committee in 1997. His description of Cargill's cotton project in Tanzania is strikingly similar to what I was told by Cargill Poland about their wet-milling operation in that country.

When Cargill first bought into the Lalago cotton project in Tanzania, Micek said: 'Farmers were being paid for their cotton in promissory notes, pieces of paper that entitled them to payment sometime in the future.' Cargill set up cotton-buying stations, established cotton ginneries that provided jobs off the farms, and started paying cash. Cargill also made money off the Lalago project according to Micek. In 1996 the company processed 30,000 tons of cotton with a value in excess of $17 million. In addition, he explained, the 18,000 metric tons of cottonseed separated from the cotton lint was sold to a locally owned oilseed-crushing business in Tanzania. The crushing business sells the oil locally and the by-product, cottonseed cake, is exported to South Africa. Micek had some advice for Africans: 'For a country to be attractive to agribusiness does not require huge asset investment, but it does require some investment.' And in the case of Tanzania, as well as other sub-Saharan African nations, he explained, this means investment in transportation. 'If the farmer can get the cotton to the buying station and the buyer can get the cotton to the gin, and the ginnery can get the cotton to the port, the cotton can become a world-priced product, generating cash for more internal investment and foreign exchange earnings for the national account', he explained. 'This moves the cycle of improvement upward.'[88]

Peanuts

The lowly peanut (groundnut) does not attract much public attention, but Cargill's involvement in the US peanut-trading and processing industry is not really surprising when one considers the size of the industry. The US has ranked third behind India and China in peanut production for many years. Per capita consumption of peanuts in the US is among the world's highest, and over half of US production goes into domestic edible uses. A quarter goes into export edible markets, accounting for more than one-third of world peanut trade. The remaining quarter is crushed for oil and meal, or used for seed and animal feed.

Cargill's position in US peanuts is nevertheless highly ambiguous. The company is a bit shrill in its advocacy of the free market and its condemnation of government interference and regulation, yet peanuts (like sugar) are among the most controlled food products grown in the US. This is possible because peanuts are a controlled crop in the US and grown under a quota and two-price system. 'Quota' peanuts are sold on the domestic market at about $700 per ton, with 'additional' peanuts sold into export at about half the domestic price. The industry has been protected for years and continues to be under Section 13 of NAFTA.

To find out more about peanuts and Cargill, I sought out Cargill in southeast Georgia. 'Sylvania Peanut Co. – Cargill Inc.', the sign on the front of the little office building said. Across the road was an abandoned traditional sharecropper's house – the long front roof covering a full-width front porch. Sylvania Peanut Company has been around a long time, buying, grading and selling peanuts, and was Cargill's beachhead in the peanut business. Following its customary strategy, once Cargill had learned something about the business it made a decision about whether to get in deeper or to make a strategic retreat. Apparently Cargill liked what it found and decided to advance, so it bought Stevens Industries in Dawson, Georgia, and rolled the Sylvania company into it. Stevens is a much larger company that processes peanuts and makes peanut butter, though it does not have a refinery for producing peanut oil – yet.

Cargill now states explicitly that it supplies peanut butter to the US school lunch programme. (Nothing like a government contract to give free enterprise a boost!) A lot of Stevens/Cargill peanut butter is sold to Procter & Gamble and marketed under one of its names, another one of Cargill's invisible presences at the retail level.

Malting

Even more invisible is the barley malt that made your beer. In 1991, before it purchased a majority interest in Ladish Malting Co., the largest malting company in the US, Cargill was already a major producer of barley malt in Europe with malting operations in France, Holland, Belgium and Spain. The purchase of Ladish moved Cargill into first place with 8 per cent of the world market, surpassing Canada Malting Co. Ltd which had been the world leader.

In 1997, Schreier Malting Co., 51 per cent owner of Prairie Malt of Biggar, Saskatchewan, sold its share of the family business, founded in 1856, to Cargill. Saskatchewan Wheat Pool retained its 42.4 per cent share of Prairie Malt. The deal gave Cargill a 45.3 per cent controlling interest in CUC Malt Ltd in Nanjing, China, and a production facility in Wisconsin.

Cargill consolidated its malting operations in the Americas – the Ladish facilities in Wisconsin and North Dakota, the former Schreier facility in Wisconsin, its interest in Prairie Malt and malting operations in Bahia Blanca, Argentina – under the name Cargill Malt in 2000. Cargill also has malting plants in Belgium, France, Germany, the Netherlands and Spain.

7 Processing: Oilseeds, Soybeans, Corn and Wheat

From time to time Cargill gets carried away with its own success. Some years ago, just before its sales and earning took a sharp drop, the corporation allowed itself to say, in an elegant, timeless (that is, undated) brochure with a slate-grey cover bearing only a wordless Cargill logo in deeper grey: 'Our performance goal is to double the size of Cargill's business every five to seven years.' How did Cargill propose to double its business? At that point it may have had the financial markets in mind, but more likely it was what the company saw ahead for the very material business of processing grains, oilseeds and palm oil.

As its figures for the first half of fiscal 2002 show (June 1 to November 31, 2002) the severe stock market decline in the fall of 2001 did not hurt Cargill. On the contrary, the company did very well. We will, after all, continue to eat – even though we have done it once does not mean we will not do it again. This is how Cargill makes money:

> Cargill earned just $351 million in 1991, a paltry return of less than 1 per cent on sales ... Not to worry, says chief financial officer Robert Lumpkins: Commodities companies make money by rapidly turning over their inventories – 15 times per year in Cargill's case. Also, the company's conservative accounting practices – including accelerated depreciation of assets – make the bottom line unnaturally thin. This year, predicts Lumpkins, Cargill will generate about $1.1 billion in cash flow, most of which will be reinvested or used for acquisitions.[89]

Oilseeds

Oilseeds refers here primarily to soy, though technically the term refers to rape/canola, sunflower, cottonseed, flax and a wide variety of crops that produce edible vegetable oils. It is difficult to figure out how to organize material on a subject as complex as processing that starts primarily with three major crops and turns them into a multitude of products. Here I try to give an historical and geo-graphical account of Cargill's development in the processing of soybeans, corn and wheat.

For many decades after its establishment, Cargill remained, as one of its brochures put it: 'a regional grain merchandiser'. This description hid the fact that the company was already evolving into a trader in commodity abstractions such as contracts, futures, and now, derivatives. More substantially, however, Cargill has always understood that the key to long-term success in trading is the capacity to store and deliver commodities. The ability to hold staple crops and other commodities off the market while awaiting, or bargaining for, a better price is an obvious way to increase one's profit with only a small risk involved. The ability to originate (gather from the farm level), store and deliver provides both leverage and another source of profit.

It was nearly 80 years before Cargill moved 'downstream' from these activities into commodity processing. In 1943 Cargill purchased three soybean-processing plants, two in Iowa and one in Illinois, and while there is little available record of the company's steady growth in oilseed processing, we do know that it acquired six mills in the US from Ralston Purina in 1985, two oilseed-processing mills and three rice-processing mills in Malawi and two palm-oil refineries in Malaysia, and by 1991 the company was operating more than 40 oilseed-processing plants around the world. The next year Cargill was supervising the building and operation of a cottonseed-processing plant in Shandong Province, China, apparently the outcome of a joint-venture agreement with the government of China. It also acquired, from Continental Grain Co., four more oilseed-crushing plants: one in Alabama; one in Argentina; and two in Australia. (Cargill also picked up three feed mills in Australia from Continental Grain in the deal.) Around the same time it purchased Huilerie Felix Marchand SA, a 145-year-old oil refiner in western France which processed sunflower, rape, soya, peanut, grape, corn and palm oil.

Cargill claimed in 1992 that Germany was the only major oilseed-producing country in the world in which it did not have an oilseed-processing facility, but it remedied that by constructing an $80 million oilseed-processing and malt plant in the port of Salzgitter, near the old East–West German border. The company was apparently unaware that Canada, where it did not have any oilseed-processing facilities, was not part of the US. It subsequently remedied this oversight.

Cargill Ltd (Canada) long had a garrulous vice-president, Dick Dawson, who once told me his version of Cargill in Canada. Cargill wanted to move beyond trading and into the originating business back in 1972, Dawson told me, and rapeseed/canola was the only major crop not under the control of the Canadian Wheat Board. So Cargill made its first investment in western Canada building a rapeseed processing terminal at North Battleford, Saskatchewan. It intended to build a crushing mill as well, but never got around to it since the terminal was sufficient to give it fair access to the rapeseed trade.

Years later, in 1994, Cargill Ltd announced that it intended to build a canola-processing plant and refinery in western Canada that would be two to three times the size of any existing plant in western Canada and bigger than either of the two big plants (ADM and Central Soy/Canamera) in Ontario that crush both soybeans and canola. Cargill cagily made the announcement without specifying the location, an unmistakable invitation to the provincial governments of Saskatchewan, Alberta and Manitoba to court Cargill with promises of favours and finance. This would give 'The Big Green Crusher', as one industry wag described Cargill, the opportunity to locate its plant according to the dowry promised – at public expense, of course. Saskatchewan won (or lost) and the plant was built at Clavet, not far from Saskatoon with the aid of $3.9 million from the New Democratic Party government of Saskatchewan.

In 1995 Cargill Canada purchased InterMountain Canola from DuPont. InterMountain produces specialty canola and rape seed, contracts with growers to grow the seed and processes the seed and markets a proprietary low-linolenic-acid canola oil under the Clear Valley brand and high-eurucic-acid rapeseed oil.

With the completion of a new oilseed-crushing facility in Australia in 1997, Cargill was able to process over 1 million tons of Australian canola, cottonseed, sunflower seed and soybeans annually. The oil is used in margarine, salad dressings, frying, baking, and by food processors and restaurants in Australia and Asia and the oilseed meal goes into livestock feed.

Cargill's growing role in Russia is described briefly on its website. Cargill opened its first Russian office, in Moscow, in 1991. By 1997 the company had 1,300 employees in Russia. The 70 employees in the Moscow office were 'involved in business activities in petroleum, frozen concentrated orange juice, vegetable oil, sugar, dairy and other agricultural products'. Cargill's major field activities are in the Caucasus. 'Cargill sales teams go into the fields to sell hybrid seeds and fertilizer, provide agronomic consulting and buy grain or barter fuel ... We're the only Western company which offers a total package for agriculture – from supplying seed to processing farmers' crops ... There are better times ahead for Russia's 150 million people. Cargill is in a good position to be part of that future.'[90]

In 1994 Cargill had begun to buy shares in a maize-processing plant in Effremov (or Efremov, or Yefremov – the spelling changes from press release to press release) about 380 km south-east of Moscow, and by 1996 it was the majority shareholder. As Cargill explained: 'With products from Mars [as in candy bars] and other local and Western manufacturers fuelling the demand for higher quality confectionery, soft drink and bakery products in Russia, acquiring a glucose plant enables us to serve customer needs for quality glucose sweetener faster than building a new facility.' Plant manager Gerrit Hueting (whom I later met while visiting Cargill Poland's office in 2001 – he is now manager of Cargill's wheat wet-milling plant in Wroclaw – see below) told how the plant brings significant benefits to Russian farmers: 'Our work with corn growers in providing technical services and crop products has also enabled these farmers to significantly increase their crop yields.'[91] Substitute wheat for corn and Polish for Russian and you have exactly what Hueting told me in Warsaw five years later!

Cargill shut the plant down in 1998 as a result of the 'financial turmoil' in Russia. Cargill says that it invested $40 million in the plant and that it would have been cheaper to build a new plant from scratch, but Cargill 'wanted to be in the market quickly', said Hueting. 'It's a learning experience. We learned possibly more than we would have ever imagined.' While Cargill started with a minority stake, it wanted a controlling interest, so it purchased shares held by employees as a result of the privatization programme. It hired a local firm to buy the shares directly from the workers, and to this day it is a source of resentment. 'They cheated us', said one worker. 'Cargill declined to respond directly to a question about the tactics used to buy shares from the workers.'[92]

In Venezuela, as a result of building the largest and most modern palm-oil-processing plant in Valencia, Cargill now has 25 per cent to 40 per cent of the Venezuelan industrial and edible palm-oil market, depending on how much is oil being imported from Colombia.

There are ongoing talks with the government about how to keep out Columbian oil in the name of self-sufficiency for Venezuela. This is a rather typical example of how Cargill portrays itself, and is accepted, as a domestic corporation acting in the best interests of the country within which it is operating.

Cargill says it started trading in Venezuela in the 1950s, but Cargill de Venezuela, CA, the fully owned subsidiary of Cargill, Inc., was not established until 1986 with a pasta and flour plant in the city of Maracaibo, Zulia State. Today Cargill de Venezuela operates in 15 locations with businesses including two pasta plants, two oilseeds operations producing palm oil, a flour mill, a rice plant that processes and produces branded white rice and another that produces parboiled rice and rice flour used mostly to prepare baby food under the brand name Mimarroz. Cargill also has a pet food mill and a sugar operation.[93]

Cargill acquired Vamo Mills from Vandemoortele International NV in 1997. The agreement includes soybean-crushing, soy-protein and lecithin plants in Gent, Belgium, and liquid-oil refineries in Izegem, Belgium, and Mainze and Riesa, Germany.

By 1996 Cargill was operating 29 soybean-processing facilities in the US. The same year it built a soybean processing plant in Tula, Mexico, adjacent to its Atitalaquia Corn Syrup Distribution Centre north of Mexico City. Cargill planned to sell soybean oil to Mexican refiners while the meal would go to the feed industry. Guillaume Bastiens, president of Cargill's Food Sector, called the project 'an excellent example of how Cargill, through NAFTA, can help increase export opportunities for US grains and oilseeds and contribute to better economic and environmental conditions in Mexico'.[94]

Summer, 1997: as the soybeans were ripening in the fields of North America, Cargill announced that it was importing Brazilian soybeans. Cargill explained that US soybean stocks were scarce and too expensive, forcing the soybean-processing industry to idle about one-third of its capacity and making it necessary for the company to import Brazilian beans to fulfil contracts with dairy, poultry and pig farmers in the US. The most immediate reaction was a 30 cent drop in soybean prices on the Chicago Board of Trade, which must have delighted Cargill, and may have been its objective in the first place. Cargill is a trader and processor, not a grower, so low commodity prices give it more room to manoeuver. Then there is the fact that Cargill is a major trader and processor in Brazil, which means that Cargill was probably importing from itself. The farmers in both the US and Brazil were the losers.

Poland

In November, 2001, I was quoted in a Warsaw newspaper as saying that Cargill is public enemy number one. The next day my Polish host and I took a cab to the new Cargill office in Warsaw. The ride was a little longer than expected because we first went to the office address given on Cargill Polska's website, but, as it turned out, the address, like the rest of the information on the website, was about three years old. Fortunately the receptionist could give us Cargill's current address because even the new phone book has the old address. One senses some ambiguity about the issue of Cargill's visibility – or invisibility.

Entering Cargill's real office, we found three young women in the reception area who seemed pleased to see us. They asked how they could help us and I said I was just interested in what Cargill was doing in Poland since the information on the website was old. They eagerly showed us their brand new website. We joked around for a few minutes before the managing director of wheat milling, Gerrit Hueting, appeared and introduced himself. When he found out who I was, he asked about the newspaper article and invited us into the board room to meet with himself, the senior manager responsible for animal nutrition (feed mills, that is) and their 'trader' (a Dutchman, a Frenchman and a Pole, in that order) for what turned out to be an hour-long conversation about Poland and what Cargill is doing there. The conversation was peppered with current Cargillisms, such as those in the leaflet given us by the secretaries: 'Supply Chain Management Solutions,' 'Food Application Solutions' and 'Nutrition and Health Solutions' – all the words one can read in the policy speeches of past and present Cargill presidents on the company's website. When the conversation was getting into a serious discussion of Cargill policy, the senior manager said he should state the corporate policy and did so. It was not news to me, but he felt responsible for playing his role. He said their main mission is to provide more food of higher quality and to bring in technology. Agriculture in Poland has to become more highly capitalized and more technology intensive. Our principle, Marcel said, is to help farmers to adjust to the new reality, to become more competitive and efficient. We provide innovation and solutions – it is up to the society to determine policy.

They were happy to tell me that the company's glucose plant (you can't refer to what comes from wheat as 'high-fructose corn syrup' (HFCS)) in Wroclaw contracts with farmers for their wheat and pays them a week or so after delivery of their crop to one of the independent flour mills that custom grinds the wheat. The wheat flour is then shipped to Cargill's single large mill in Wroclaw. They are proud of the fact that Cargill provides Polish farmers with a market for a crop that is essential to a rotation with sugar beets; that it has the wheat ground in Polish mills; and that it then processes the flour into glucose and fructose syrups (*syropow glukozowych i izoglukozowych*) for the Polish food industry, gluten for the Polish baking industry, and meal for Polish animal feed. Cargill Poland neither exports nor imports wheat or wheat products. (What else goes into the animal feed and what the trader trades is another question, which we did not discuss.) They did not need to tell me that their current treatment of the farmers is a very good way to build loyalty, nor did they need to point out that given the state of Polish agriculture, any payment would be better than none. We noticed later, at an 'animal nutrition' plant near Warsaw, that there is no ambiguity about Cargill's visibility in the countryside: its trucks are all clean, bright, and clearly identified: 'Cargill'.

Cargill established its eastern European activities in 1991 with the opening of a Warsaw office and implementation of its highly refined strategy of establishing its presence – a 'beachhead' – in a new region in the low-profile and apparently innocuous area of feed milling in 1992 'through a joint venture with the State Treasury'. Consistent with its principle of holding a controlling interest in any joint venture, Cargill holds 60 per cent of the new company, Cargill Pasze SA, with the remaining 40 per cent held by the provincial government of Plock. Cargill now has six feed mills and service centres throughout Poland.

Having established its presence, Cargill then moved into wet milling, building the plant near Wroclaw, 400 km south-west of Warsaw, to produce starch and glucose for Polish confectionery, ice cream and soft drinks manufacturers. It is the only mill in central and eastern Europe designed to refine glucose from wheat. The plant started production the following year and in 1999 the plant was expanded to a capacity of 120,000 tons per year with the added ability to produce fructose.

The leaflet referred to above also presents Employment Opportunities with Cargill:

> When you join Cargill, you work in a highly demanding and fulfilling environment. We believe in hiring talented and strong individuals and giving them challenging positions ... A career with Cargill can give you the intellectual challenge and job satisfaction that allow you to truly enjoy your job.

The top managers we met in the Cargill office struck me as filling that bill. You just have to accept the basic premises of corporate agribusiness and western high-tech, capital intensive agricultural production.

Corn Processing

> There are two types of milling process depending on the product desired. Dry milling is the process used to produce what we commonly refer to as flour (with other byproducts such as wheat germ), whether it is wheat or oat or barley flour, corn flour (which has become very important for tortillas), or any other grains to be used for human consumption. The first stage in the deconstruction of corn by the wet milling process is steeping (soaking) the corn to soften the kernel so that its major component parts – germ, starch, gluten and hull – can be subsequently separated. The starch becomes a slurry that can then be converted by means of enzymes into conventional corn syrups, high fructose corn syrup (HFCS), glucose, dextrose, crystaline fructose, corn gluten feed and ethyl alcohol (ethanol).

Cargill's own information about its corn wet-milling division does not, typically, provide any figures, but the company does say that it started in the wet-milling business in the US by buying a plant in Cedar Rapids, Iowa in 1967. Cargill built a second corn wet-milling plant in 1976 on President's Island in the Mississippi River at Memphis. Being on the river meant that Cargill could supply the plant with corn by barge from upstream elevators very cheaply.

A year later, in 1998, Cargill took greater advantage of river transport and its skills in specialized commodity movement and launched the first barge designed specifically to carry HFCS and other liquid sweeteners from its Memphis plant. The barge, with its six stainless-steel tanks, is one of a fleet of 14 that will allow the Memphis plant to ship sweeteners on the Mississippi River at freight rates considerably lower than those for rail shipment. In addition to the state-of-the-art barges, the $30 million delivery system includes a new loadout facility at the Cargill plant as well as receiving terminals at Tampa, Florida, and Houston, Texas. The facility has long used traditional barges to carry corn from the Midwest for processing, and to transport co-product feed ingredients from the plant to Gulf Coast export facilities. The barge project is one of a number of initiatives at the plant that have benefited from Cargill's participation in the Payment in Lieu of Taxes (PILOT) programme established by the Memphis/Shelby County Industrial Development Board to encourage business investment and job creation in the area. In exchange for real and personal property tax abatements of $8 million over a 13-year period, Cargill agreed to invest $80 million and create 28 new jobs at an average annual wage of $44,210 by December 1998.[95]

Cargill chairman Ernest Micek elaborated in a speech to the Corn Refiners Association in 2000, saying the company developed a continuous milling process and a continuous corn-syrup refining process before developing HFCS around 1980. Micek said Cargill went from a capacity of 8,000 bushels of corn ground per day in 1967 to more than a million bushels per day in 2000. A January, 2000, news story indicated that Cargill's corn wet-milling division produces, in the US, about 600,000 tons of Sweet Bran (a corn-derived cattle feed), 100,000 tons of corn gluten meal, 50,000 tons of corn oil, 1.5 billion pounds of HFCS, and 270 million litres of fuel-grade ethanol. According to one analyst, Cargill controls 21 per cent of US corn-milling capacity.[96]

As of January, 2002, the company operates corn wet-milling plants in the US in Eddyville, Iowa, Blair, Nebraska, Dayton, Ohio, Memphis, Tennessee, and Wahpeton, North Dakota. It also has a plant in the Netherlands that can utilize both maize/corn and wheat as feed stocks, a maize-based plant in England, as well as plants in Russia, Turkey, Brazil and Poland, as we have just seen.

To gain an insight into how Cargill's corn wet-milling businesses have grown, one needs to look in more detail at the development of its mid-west US operations.

In 1985 Cargill built a corn wet-milling complex at Eddyville to produce HFCS. Five years later Cargill added a $45 million citric-acid plant to transform corn-derived liquid dextrose into citric acid. By the end of 1992 the plant was producing 36,000 tonnes of citric acid per year and supplying 20 per cent of the US market. Another addition to the plant made it possible also to produce 15 million pounds annually of sodium citrate, the sodium salt of citric acid. Both citric acid and sodium citrate are used in carbonated beverages, and sodium citrate is used to suppress the bitter aftertaste of the saccharin used to sweeten many low-calorie beverages. Both products are also used as biodegradable alternatives to phosphate in detergents. The next addition to the Eddyville complex was a $30 million ethanol refinery.

Cargill's interest in ethanol goes back to the 1970s, but the technology for ethanol was undeveloped then and it cost more to produce ethanol than it was worth as a fuel component. The industry was, and still is, dependent on government subsidies. Although the economics of production may not have changed significantly, the politics have; US legislation mandated the use, starting in 1995, of ethanol as a blend in gasoline in order to make it burn more cleanly. Tax incentives have been used to encourage the production of ethanol and Cargill's rival, ADM, has been the noisiest lobbyist both for its use and for the tax incentives and/or subsidies required to make it attractive.

After Eddyville came Blair. In 1995, when Cargill opened its $200 million corn wet-milling plant at Blair, it announced that it had already begun a $97 million expansion of the plant, which produces fuel-grade ethanol and cattle feed in addition to HFCS. Next they added a $36 million plant to process corn germ into corn oil. A Cargill handout suggested that when the Blair plant was fully operational Cargill would be buying and processing approximately one of every 30 bushels of corn grown by the American farmer.

In 1997 the plant was expanded for the production of erythritol, 'a sugar alcohol derived from corn that has 70% of the sweetness of cane sugar and only 0.2 calories per gram'. Erythritol was developed in Japan through a fermentation process using dextrose made from corn. The new $50 million plant was a joint venture between Cargill and Mitsubishi Chemical, which had a patented process for producing erythritol. The new project brought total capital investment in the Blair complex to about $400 million. (Mitsubishi Chemical has since dissolved the joint venture but maintained Japanese erythritol marketing operations.)

A lactic-acid plant was also added to the complex in 1997. Lactic acid is a natural organic acid used as a flavour agent, preservative and acidity adjuster in foods. This was a joint project of Cargill and CSM n.v. of Amsterdam. The new plant also supplied polylactic-acid (PLA) polymers to Cargill's EcoPLA plant near Minneapolis, which was really a pilot plant for Cargill Dow Polymers LLC, 50/50 limited liability company formed by Cargill and Dow Chemical to develop and market PLA polymers. PLA resins are composed of chains of lactic acid that can be produced by converting starch (from corn or sugar beets) into sugar and then fermenting it to yield lactic acid. Water is then removed to form lactide, which is converted into PLA resins using a solvent-free polymerization.

In the late 1980s a lot of excitement had been generated by the corn industry over the advent of a 'biodegradable' plastic that could be used for all kinds of purposes including garbage bags. The novelty was the addition of corn starch to the normal plastic. When exposed to the weather, the corn starch would dissolve and the plastic would crumble. But that was all it did. There was, in fact, just as much plastic left, but it was in little bits. Cargill-Dow's new PLA product is based on a totally different process and is, they claim, completely biodegradable (and there is no reason it shouldn't be, but whether producing plastic from corn is ecologically sound is another matter).

In January, 2000, Cargill Dow Polymers announced that it would build a $300 million 'world-scale facility' at the site of Cargill's corn wet-milling plant at Blair to manufacture PLA, utilizing natural plant sugars from corn to make proprietary PLA polymers, which the company has named NatureWorks PLA. According to Cargill, the polymers can be made into utensils, packaging or fibres for cloth and carpeting and is expected to provide competition to cellophane for packaging due to its complete biodegradability and low price. Corn costs less than wood pulp as a feedstock and the manufacturing process is less complex, with the result that the polymer is cheaper to produce. The company is working on extending the technology to be able to use other crops, such as wheat and sugar beets, and agricultural waste for feedstock. The plant was due to come on stream in late 2001.

Yet another addition to the Blair complex was Midwest Lysine, a $100-million production facility that opened in 2000. This is a joint venture of Cargill and Degussa-Huls Corp., a subsidiary of Degussa-Huls AG of Germany. Midwest Lysine manufactures a premium lysine amino acid which is used in livestock feeds. The dextrose that Midwest Lysine utilizes as its primary feedstock is a primary product of Cargill's corn wet-milling plant next door. This brought Cargill's total investment in Blair to more than $400 million.

In his 2000 speech to the Corn Refiners, Ernest Micek explained that one of the reasons he was optimistic about the corn-milling industry was that its products came from a renewable resource. He said we would be hearing a lot more about 'eco-efficiency' in the future, which he defined as 'the creation of economic value while reducing environmental impact and resource use'. What Micek did not mention was that industrial corn production is a very heavy user of non-renewable energy in the form of petroleum and natural gas products (diesel fuel and nitrogen fertilizer) and mined fertilizers (phosphate and potash). He did not mention that Cargill is a producer/supplier of fertilizer. Nor did he mention the mammoth public subsidies (providing about half of farm income) going to the big industrial farmers in the US to keep them producing cheap feedstock for Cargill's processing.

Cargill moved yet further downstream by forming a partnership with Hoffmann-La Roche Ltd of Switzerland to build a plant to man-ufacture natural-source vitamin E as part of Cargill's Eddyville, Iowa, corn-processing complex. Cargill operates the plant and Hoffman-La Roche is responsible for marketing. The plant utilizes technology developed by the two companies to extract vitamin E from a product of soybean-oil refining.

Cargill has also built an itaconic acid facility at its Blair corn-processing complex, having purchased the itaconic acid business of Cultor Food Science of Finland. Cargill is now the world's largest supplier of itaconic acid, which is used in everything from the latex backing on carpets to coatings on paper that make water bead up.

Micek was not simply using a simile when he said, in his speech to the Corn Refiners, that while corn milling itself had not changed much in 30-some years, the 'back end' now 'looks a lot more like a large-volume drug store'. He gave a simple explanation for the more complex product line: 'Regular glucose is about 8 cents per pound. Fructose is 12 cents per pound. Citric acid is 70 cents per pound. And itaconic acid is $1.80 per pound.'[97]

Cerestar Acquisition

Cargill made a big leap in October, 2001, with the announcement that it intended to acquire 56 per cent of Cerestar from Montedison. Cerestar is one of four companies formed through the de-merger of

Eridania Beghin-Say, the agrifood company of the Montedison group of Italy, earlier in 2001. Cereol, another of the four companies, is a global oilseed-processing company that includes Central Soya as its North America division. French law requires Cargill to make a tender offer for the remaining 44 per cent of Cerestar shares that are held by the public.

Cerestar, with 16 production facilities in ten countries, has a global product line – glucose, high-fructose syrups, and animal-feed ingredients from both wheat and corn – identical to that of Cargill. It controls 30 per cent of Europe's market and 5 per cent of the North American market for these products. The addition of Cerestar's three HFCS plants in the US will effectively increase Cargill's share of US total capacity to 30 per cent, placing it on a par with ADM, and reinforce its position as the leading supplier of glucose and dextrose to the North American market. The deal will also give Cargill access to isoglucose (HFS) production quotas in the European Union.

At the end of October, 2001, Standard & Poor's (S&P) and Moody's both reduced their credit ratings for Cargill, triggered by the announcement of the purchase which is figured to cost a total of $1.1 billion, including assumed debt of an estimated $364 million. The rating agencies figure the total debt of Cargill and Cargill's Employee Stock Ownership Trust at about $4.5 billion. However, they also say that Cargill's business profile will be enhanced by the acquisition of Cerestar's starch business since it will strengthen Cargill's position in the European market and add Cerestar's higher margin, value-added product portfolio to its strong but commodity-oriented product portfolio. The S&P report also commented: 'Cargill's product and geographic diversity, as well as its vast communication and transportation network, helps optimize commodity movements and provides competitive advantages. There is also less exposure to any one segment of US agriculture or to any one international locale.'[98] In January, 2002, Moody's rated Cargill A1 among the companies in the food world that have 'effective business models that create shareholder wealth'.[99]

Dry Milling

Wheat Flour

Cargill has long been a major wheat-flour miller in many countries, including the US. For a while, it was even cited as the largest flour miller in the world. This was in 1982 after it bought Seaboard Allied

Milling Corp, the seventh largest flour miller in the US, from Seaboard Corporation. Cargill is now the third largest flour miller in the US behind ConAgra and ADM with 18 flour mills in the US. Its largest mill, in Albany, New York, has a capacity of 1,000 tonnes per 24 hours. In 1993 the company closed the 96-year-old flour mill in Buffalo where it got its start in flour milling in 1981.

In what *Milling & Baking News* described as 'the largest flour mill closing in decades', Cargill announced the shut down of three of its US flour mills in April, 2001. Two of the three were to be dismantled, the third 'mothballed'. The current round of mill closings was the largest since 1995 when General Mills announced the closing of nine of its 17 mills, including its very large, but venerable, Buffalo, New York, mill which had the same capacity as the three mills closed by Cargill. ADM has also closed three smaller mills, and together, all the mills closed accounted for 5 per cent of the total daily milling capacity in the US.

While the flour millers may open and close flour mills like playing an accordion, farmers do not have the same flexibility. A retired general manager for Cargill Flour Milling candidly said:

> Growers ... have no say over the prices they receive, nor do they have any way to negotiate the pass-through of their costs to consumers. They sell at prices set by others on the commodity market. Growers cannot smoothly adjust production in response to supply and demand like millers can ... In addition, there are 300,000 wheat farmers in the US and there is no way for that number of individuals to make smooth adjustments in response to supply and demand ... In contrast, look at the concentration in the milling industry which buys the wheat.[100]

A sign of the increasing cooperation between erstwhile competitors is the 2001 agreement between Cargill and CHS Cooperatives, a producer-to-consumer cooperative system operating in the mid-west and west of the US, to form a limited liability company, Horizon Milling, LLC, to operate the US flour-milling businesses of both companies with Cargill as the managing partner. The partnership includes 16 Cargill flour mills and five Harvest States mills, with a flour-milling capacity about equal to that of ADM. CHS Cooperatives itself is the amalgamation of Cenex and Harvest States cooperatives; Harvest States is the name of its grains and foods division. The name of the new joint venture, Horizon Milling, was chosen to reflect 'a commitment to expand the horizon of opportunities for flour customers with a national scope of quality, consistency and product innovation', according to the press release. There was no mention of any benefits for farmers.

In announcing plans for the venture some months earlier, CHS said it was at a crossroads: 'too big to be a niche player but not big enough to compete successfully in a consolidating industry'. While acknowledging that even some of its members questioned the partnership of a grower-owned cooperative and the world's largest privately held company, senior CHS management said: 'We have a lot in common even if we do come from a different ownership base.'[101]

Earlier in 2001 Cargill had fired 80 workers and closed three of its 19 US flour mills after excess production by Cargill, ConAgra Foods and other grain processors depressed prices, and ADM Milling discontinued operations at its flour mill in Destrehan, Louisiana, 'due to excess industry capacity and poor economic conditions'. ADM shut down its 85-year-old flour mill in Des Moines, Iowa, in January, 2002. (ADM has over 22,000 employees, 275 processing plants and net sales for the fiscal year ended June 30, 2001, of $20.1 billion.)[102]

Cargill moved into rice milling in 1992 with the purchase of the assets of Comet Rice Mill in Greenville, Mississippi, from Prudential Insurance Co. The Comet mill is the biggest in Mississippi and is located on the river in the heart of the delta. Its barge-loading capacity gives it a distinct advantage in rice exports, since most of the 30 or so rice mills in the US are landlocked.

Corn

After gaining some practical experience in dry milling corn in a small mixed-product mill in Saginaw, Texas, which it closed in 1993, Cargill moved into full-scale corn dry milling in a joint venture with corn dry miller Illinois Cereal Mills Inc. The joint venture was named Illinois Cereal Mills Ltd. A year later Cargill joined with Kellogg Co. to transform Seaforth Corn Mills, of Liverpool, England, into a joint venture between Illinois Cereal Mills Ltd and Kellogg Co., which had previously been the sole owner of Seaforth, the largest corn dry miller in Europe. The third step was for Cargill to acquire 100 per cent of Illinois Cereal Mills Inc., which had previously been an employee-owned company. The result is that Cargill is now one of the two largest corn dry millers in the world. Remember that when you are eating your tacos or tortilla chips.

8 Invisible Commodities

It used to be that companies were known by the products they bought, sold or manufactured. This is still true of Cargill, but along with the growth of its real commodity production, handling and processing activities there has been a steadily growing segment or layer of the company's business that is not only invisible to the vast majority of people around the world but which is increasingly abstract or even non-existent. This aspect of Cargill's financial activity has become the second highest contributor to the company's overall economic performance.

Cargill shuns the word 'speculative' – at least in public – while insisting that its financial management is conservative. As if to prove the point, the term 'risk management' is used to describe the invisible transactions of its FMD and its other financial activities.

The basic mechanism of risk management in commodities trading, from the farmer on up, is hedging, which Cargill defines as 'the process of transferring price risk from someone who does not want such risk to someone willing to accept it'. As *Forbes* magazine described it: 'Every time it buys a commodity, it hedges by selling a contract to offer a like amount at some future time. When it sells the commodity, it buys the futures contract back. It is a highly conservative and safe approach from which Cargill never strays.'[103]

Thus when Cargill buys a carload of real barley at price x, it sells a contract to deliver a like amount of barley at the same, or slightly higher, price at a specified future date. That is, the contract for future delivery of the commodity barley itself becomes a commodity that can be bought and sold. Whatever the fate of the real barley, Cargill knows that it can actually sell that amount of barley at a known price in the future. (Cargill has long known the value of always having access to real grain in storage.) If, in the meanwhile, it can get a higher price for the real barley, it will sell it, perhaps at the same time redeeming (buying back) the futures contract that it had purchased earlier or purchasing more barley to maintain its position. (The commodity barley and the commodity contract-for-delivery-of-barley become equivalent – or equivalently abstract.)

Conversely, if Cargill decides to sell oats that it has in storage to Saudi Arabia because the price is right or the US government is offering a subsidy, at the same time it will buy a contract to take delivery of a similar amount of oats somewhere in the world at a future date for a price the same as or lower than that of the sale just made, if possible. This way it keeps itself covered and avoids undue risk at the cost of passing up undue speculative profits. The extent to which it actually follows its own advice is another issue, as well as what speculative financial activities are taking place elsewhere in the corporate labyrinth.

The company is avid in promoting its expertise in risk management to farmers, telling them that:

> When it comes to marketing your crop, you can count on Cargill's experience to help you get the best return ... We offer flexible grain marketing alternatives to reduce price risk and boost profits. If you prefer, we can store your crop to help reduce your on-farm investment, and position your grain for sale.[104]

One should not misunderstand such an expression of concern for the financial welfare of farmers. It is Cargill's task to make money coming and going, that is, buying as well as selling. It positions itself in the middle while presenting itself as the farmer's business associate, if not friend. The profit opportunities resulting from such a strategy would seem to be boundless, if prudently taken in small increments.

It has been sad to note over the past decade the removal, in both Canada and the US, of extension agents employed by provincial and state governments to provide information and advice to farmers. These publicly employed agents and 'ag reps', as they were referred to in Canada, might have had an ideological commitment to production agriculture, but they did not have a financial interest in the farmer's business dealings, for better or worse. Now virtually all ag consulting is on behalf of some corporation, be it a fertilizer company, a seed company, a drug-biotech company, or a corporation with an interest in selling all of these. Or buying the crop. Or all of the above. So the question must always be asked (though it seldom is) whose interests are really being served by the advice given?

Historically, within the capitalist sector of economic activity, trading has moved from the face-to-face buy–sell relations that most people experience in their day-to-day shopping – from village market to superstore – to trading real commodities by contract on specifications and then to trading in contracts for not-yet-existent commodities (futures). The latest level of abstraction is the 'derivative', which is one form of what are now referred to as 'financial instruments' or 'financial devices'. According to *Fortune* magazine: 'Derivatives are financial instruments whose value is tied to something else – called an "underlying" by the trade – such as equity or an indicator like interest rates.'[105]

The *Fortune* magazine article quoted above provides a very good illustration of a derivative, in this case an option, by using a story by Aristotle about a man who was gifted at reading the stars: 'Thales foresaw an abundant olive crop and used some of what little money he had to reserve exclusive use of all the local olive presses. Essentially, Thales was buying options, with the underlying being the rental rate for the presses.' When the crop was harvested, thanks to his monopoly, Thales could charge handsome fees for the use of the presses, paying out the much lower rental to the owners of the presses as negotiated earlier in the season. His only additional expense was the cost of the option itself. The contracts that are traded as derivatives are all similar to Thales' option on the use of the olive presses: 'A change in the value of the "underlying" benefits one party to the detriment of the other. The increased value of Thales' olive press option, for example, was squeezed out of the press owners, who missed the chance to charge more when demand increased, and local farmers, who had to deal with Thales' monopoly rather than competing owners of the presses.'[106]

A derivative, apparently, can be derived from anything: interest rate movements, currency exchange rates, stock market indices or mortgages.

Hybrid (which in agriculture usually refers to a crossbred plant from two distinct parent lines) is an appropriate term in this context, since Cargill's most rewarding line of business is based on a slightly different form of hybrid, corn itself. There are hundreds and hundreds of varieties of hybrid corn on the market for farmers to choose from, but there is little real genetic difference between them. Each is more esoteric in its supposed differentiation than the one it replaced, just like the financial derivative. The simile can be taken further: a brokerage like Goldman Sachs, or a company like Cargill, can create their own hybrids. Cargill is, at one location, developing new hybrid lines of corn while, at another location, developing equally esoteric hybrids of derivatives quite possibly based on its global trade in corn. The commodity trade in corn itself also involves dealing in currency exchange rates, interest rates and the shifting prices of the products to which corn contributes. All of this, of course, has little to do with food, despite Cargill's presentation of itself as a food company.

As wheat prices rose in the latter half of 1994, there was a reasonable inclination to attribute the rising prices to the severe drought in Australia. Experienced voices, such as the trade journal *Milling & Baking News*, however, began to wonder if the 'perplexing price moves' might be related more to 'the needs of the fund business' than to the needs of the food business, referring to the trading activities of derivative-based mutual funds. *Milling & Baking News'* concern focused on derivatives that are meant 'to allow investors to take financial positions in commodity markets without the physical exposure', that is, without ever actually owning any real, physical commodities. 'After all,' *Milling & Baking News* editorialized, 'everything about traditional markets, including one of the first derivatives, grain futures, is meant to maintain a relationship between these markets and "physical exposure" through the delivery system'.[107]

The evolution from buying and selling real grain to trading in futures and finally derivatives is both subtle and continuous, and the largely invisible character of futures and other financial instruments means there are few discernable landmarks in the development of financial markets. It is therefore not surprising that there is no available record as to when Cargill really began its worldwide trading activities, activities which were, by their very nature, because of the time and distance involved, speculative.

Some accounts say that Cargill opened an office in New York for offshore trading in 1922 and an office in Argentina in 1929, while other accounts report that it established its first trading office outside of the US in 1954 in Panama. Cargill did establish a Panama office in 1954 and initiated a global offensive in grain trading, aided by the passage that year of PL 480, the US Food for Peace legislation that continues, almost half a century later, to subsidize the sale of US commodities abroad to the benefit of US agribusiness, particularly the transnational traders (see the previous discussion of PL 480 in Chapter 4). It is not possible to believe that it was all mere coincidence. (The proposed 2002–3 US budget for international programme funding, including PL 480, is around $6.5 billion.)

The Panama office did not last the year (except as a tax-avoidance facility), as it very soon became obvious that communications facilities there were inadequate to support international trading activities. A new trading company was established in Winnipeg, but this too proved to be the wrong location and the business was moved to Montreal and renamed Kerrgill Company Ltd.[108] The name was coined from Kerr-Gifford and Cargill, indicating the genesis of the business, but was soon changed to Tradax Canada Ltd to reflect the non-specific geography of the new company's activities. That is, 'Canada' indicated its base of operations while 'Tradax' implied some sort of trading.

In 1956, while it was still located in Montreal, Tradax acquired Andrew Weir (Far East) Ltd, an importer and supplier of foodstuffs that had been set up just after the war and which had been acting as agent for Tradax. This provided a receiver for the commodities shipped out of Cargill's Kerr-Gifford facilities in Oregon, which Cargill had purchased in 1953. The purchase of Kerr-Gifford provided Cargill with terminal facilities in the Vancouver area as well as the north-western US.

A decade later, in 1966, the Tradax office was moved to Geneva, Switzerland, where it may remain, although the name no longer appears in Cargill publications. *Business Week* wrote in 1979 that Tradax was located in Panama but that this was merely a 'letter box' tax shelter. If Tradax has vanished, or been absorbed, or been 'disappeared' to serve some devious purpose, there are lots of other departments, divisions and subsidiaries to take its place with names such as FMD, Cargill Investor Services Inc. (CIS), Cargill Financial Services Corp, Access Financial and Cargill Global Funding plc, names which appear only in investor research reports like Dan Bosworth, Bloomberg or S&P.

It may be that only a small handful of people actually know the true structure of Cargill's financial activities and businesses, but it is not necessary to understand the structural intricacies to understand their functioning. The goal, of course, is to accumulate capital, and the more mechanisms one can invent and control for doing so, the greater the chances of success. By shifting expenses and profits from place to place, while also trading in both real nonexistent commodities such as futures and derivatives, Cargill, or any other transnational, can mystify even the best government auditors. What is given as Cargill's sales or profits for any one aspect of its business, or any national unit, like Cargill Ltd of Winnipeg, are what Cargill chooses to allocate to those categories for its own purposes. If there is a tax advantage in showing a loss somewhere, that can easily be arranged. By the same token, profits can be syphoned off through Tradax or some other division in one or another location sight unseen, or made to appear to suit corporate plans. What to one accountant may appear as a 'dividend' could just as easily appear to another as a 'management fee' or an inflated price for goods or services, particularly if the accountant in question worked for the corporation being analysed or simply shared the same ideological orientation.

The Financial Markets Division

> This business is built on the recognition that money is the ultimate commodity ... a commodity that can be traded, processed and managed, subject to the same rules and risks found in other commodity markets ... Financial assets are repackaged and redistributed to add value to the products.[109]

The origins of the FMD go back to 1973 when Cargill Leasing was established to take advantage of changes in US tax laws that made it advantageous for companies to lease a wide range of equipment rather than purchasing it. This advantage exists even if the owning and leasing arrangements are carried out between subsidiaries within the same company. From these beginnings, financial markets and financial services activities have become a major contributor to the company's income.

In 1994, the leasing division owned and leased machinery, trucks, rail cars, computers and real estate and other assets – including more than 1,500 trucks and 1,650 trailers, aircraft (leased to United Airlines and others), locomotives and poultry-processing equipment. In 1998, reflecting a shifting corporate strategy, Cargill sold its truck, machinery and computer-leasing company (with assets of approximately $625 million) to Milwaukee banking company Firstar Corp. At the time, Cargill said the sale marked the last substantial move to refocus the company's financial services division on the buying and selling of financial instruments that help it manage risks in its international agricultural operations.

As reported in *Corporate Report Minnesota* in 1994: 'Cargill has 128 years of collective wisdom on how to manage risk. The proof: long-term debt at the end of fiscal 1993 amounted to just 29% of Cargill's $6.144 billion in total capital. A rock-solid balance sheet allows the FMD to borrow more cheaply than banks and to make trades banks cannot.'

The company trades with Wall Street firms, rather than competing against them, because it considers it cheaper to 'rent distribution' from them than to build its own network of sales people. In terms of sourcing and intelligence gathering, on the other hand, 'few companies of any type can match Cargill's international presence, with more than 800 offices in 60 countries'. Intelligence-gathering is a function of any active trading office, of course, but some of Cargill's offices around the world that are identified as trading offices are simply intelligence-gathering locations. And then there are those consultants advising farmers and selling feed that report crop conditions to Cargill via their laptop computers.

The Emerging Markets Department is, as the name implies, a line of business taking advantage of the particular opportunities found in the unstable economies of the third world. The Emerging Markets Department attributes its growth 'to its ability to identify trends, find solutions to market barriers, creatively develop trading positions and capitalize on inefficient markets'.[110]

Cargill Investor Services (CIS), founded in 1972 and one of Cargill's more invisible lines of business, was designed to capitalize on the communications and trading facilities developed for the corporation's own internal use by charging a commission on investment services provided by CIS to outside clients. The first CIS office outside the US was opened in Geneva in 1980, and in the mid-1990s there were offices in Geneva, Hong Kong, Kansas City, London, Minneapolis, New York, Paris, Sydney, Taipei and Tokyo. 'For our customers who trade in global markets, CIS offers capabilities that are accessible 24 hours a day to handle electronic markets, EFPs, cash forex [foreign exchange] and all international trades', advised a Cargill ad. 'There are few firms who have the global reach of CIS, our policy of no proprietary trading [trading on Cargill's own account], our parent's financial strength, our breadth of market coverage.'[111]

CIS (www.cis.cargill.com) takes advantage, for profit, of the new markets emerging around the globe that bring with them 'the volatility and opportunity that result from the presence of political and economic uncertainty'.[112]

Since the mid-1990s Cargill has reflected this 'volatility and opportunity' in the ceaseless reshuffling (and reorganizing and renaming) of financial activities. For example:

- Cargill Financial Services Corp 'will provide $150 million in financing to Qualis Care LP, a New York-based health care receivable management and financing firm'. The deal is 'Cargill's first investment in the health care industry. Qualis will use the funds to finance medical receivables generated by hospitals, nursing homes and physicians. "We not only finance and purchase receivables, we have an accounts receivable management company that also takes over the back office for the provider, so we also lend against those receivables", said Qualis' chief operating officer Mike Gervais. Receivables are later sold in the capital markets as asset-backed securities to institutional investors.[113]

- Cargill Financial Services agreed, in 1995, to provide $600 million in funding over five years to ViatiCare Financial Services, a company that provides financial services to the terminally ill. The financial service it provides, called a viatical settlement, allows a person with limited life expectancy to convert a life insurance policy into cash by selling or otherwise assigning it at a discount to a third party, such as ViatiCare. When the policyholder dies, the insurance company pays out the full value of the policy to the new owner of the policy. In exchange for the funding agreement, Cargill obtained a minority equity interest in the company and a position on the board of directors.[114]

- Cargill withdrew as one of the main partners of International Property Corp, the consortium that acquired Canary Wharf in London for $1.24 billion, in 1996. According to the *Financial Times*, Cargill withdrew when its proposal to increase its stake was rejected by the other consortium members.

- The financial problems of Stratosphere Corp of Las Vegas has brought to light another Cargill investment, this time in a gambling casino. Through its Value Investment group, in its FMD, Cargill owns about a third of Stratosphere's $203 million in first-mortgage notes. Cargill's Value Investment group, formed in 1987, buys distressed assets at bargain prices.[115]

- Cargill Financial Services purchased more than $100 million in bad commercial mortgages in Japan in 1997, paying $40 million for the loan package, backed primarily by office properties.[116]

- Cargill negotiated a deal with Tokyo-based Yamaichi Finance Co., an affiliate of Yamaichi Securities Co., which collapsed in November, 1997, 'under the weight of $2 billion in hidden losses'. Cargill is scaling back its involvement in consumer

lending in the US due to problems and losses in its consumer
lending units.[117]

- Cargill sold its sub-prime mortgage and mobile-home lending
  unit, Access Financial, in 1998 after the unit had caused Cargill
  to take a $90 million charge to cover the high rate of mobile-
  home defaults. With an outstanding publicly traded debt of $8
  billion, S&P's and Moody's Investors Service had threatened to
  downgrade the company's credit rating if something was not
  done about the financial services division.[118]

- Cargill suffered large losses in 1998 from gambles its financial
  unit made in Russia. At least two large positions Cargill
  Financial took in the Russian debt and foreign exchange
  markets apparently went wrong. The losses were reported to
  be roughly $150 million. In the 1994 Mexican peso crisis,
  Cargill traders bet correctly and made millions for the
  company. Gregory Page, the corporate vice-president responsible for Cargill Financial, spoke with pride of the financial
  group profitably cornering more than 25 per cent of the
  Russian bond market in its formative days.[119] Greg Page is now
  president of Cargill, next in line behind chairman and CEO
  Warren Staley. 'Regarded as a numbers-oriented taskmaster,
  Page helped Cargill quit several money-losing finance
  businesses and refocus on its commodity trading and
  processing roots.'[120]

- Cargill has a website for its Asset Investment & Finance Group
  which says: 'We are an experienced buyer of sub- and non-
  performing assets ... Over 250 worldwide transactions with a
  face value of over $5 billion.'[121]

In 1996, Cargill opened a new $4.7 million financial service centre
in Fargo, North Dakota, to consolidate basic accounting functions
for all product lines in North America. Cargill received a $500,000
loan and a $100,000 grant from the North Dakota Development
Fund to open the centre. Cargill has similar centres in the UK,
Singapore, Australia, Holland and France.

The activities pursued by Cargill's financial operations are in no
way unique, and while billed as capital formation or risk
management, when put into the larger picture of current global
finance, these activities contribute to national indebtedness and the
polarization of wealth and deprivation.

By the mid-1990s, the bond market – and the overall financial sector – had become a powerful usurper of control over economic policy previously exercised by ... elected officeholders ... By the 1990s, through a 24-hour-a-day cascade of electronic hedging and speculating, the financial sector had swollen to an annual volume of trading thirty or forty times greater than the dollar turnover of the 'real economy'.[122]

It is difficult to comprehend the ramifications of this electronic, speculative trading. In fact, I dare say that no one really does. Considering Cargill alone, one has to wonder what the ratio is between the real economy of the goods and commodities traded and processed and the non-existent financial transactions in Cargill's year-end results. The real economy that defines the day-to-day options of the vast majority of people pales against the shadowy achievements of finance, but at least Cargill is still committed to the real economy, to providing real goods and services, and even to providing real wages, however limited.

9　E-commerce

Visible partnerships and alliances, or buyouts and mergers, can be expensive and might even arouse the attention of anti-combines investigators. Invisible cartels utilizing electronic 'exchanges' may be an effective way for large TNCs to reduce competition and increase profits. It may also be an effective way to get rid of small nuisance players – in a way very similar to utilizing health and other standards (such as ISO) to drive out competition from smaller players. The important point to remember is that such corporate strategies always have the objective of reducing risk (competition) and increasing profit.

> Shifting from a deal-making trade company to a world-leading processor has changed how Cargill does business. It wants to be a reliable supplier of goods and services. Toward that end, it has formed a honeycomb of partnerships and alliances with customers and other companies ... Now comes e-commerce, and Cargill is blowing away ties to the 'old economy' in which it profited from transactions as it controlled the use of resource commodities.[123]

The 'web-based' 'exchanges' described below were all announced in the three months starting on March 1, 2000, and there have been more since then. The following information was taken from www.cargill.com in 2000, and while some of it is still posted, much of it has disappeared, or has been replaced with introductory hype and a place for you to subscribe.

- Rooster.com was formed by Cargill, Cenex Harvest States Cooperatives and DuPont as 'a comprehensive Web-based marketplace to include local farm retailers, cooperatives and manufacturers'. 'Rooster.com is designed to make it easier for farmers to do business on the Internet. It will be a two-way virtual electronic mall for the agricultural industry with three initial anchor tenants [ADM and Dreyfus were added after the initial announcement] and scores of other retailers that will be open 24 hours a day, seven days a week. Here, farmers will be able to market their crops and buy fertilizer, crop protection products, other farm supplies and equipment – all through the same businesses they work with now.'

- Novopoint.com is 'an open, Internet business-to-business (B2B) exchange for food and beverage manufacturers and their suppliers' formed by Cargill and Ariba Inc. to 'give buyers and sellers of food ingredients, packaging and related services a single, convenient place to connect with each other, conduct transactions and better manage their supply chains. The company will serve participants of all sizes from every facet of the food industry, including buyers and suppliers of oil, sugar, colorings, packaging, chemicals, freight and everything in between.'

- LevelSeas.com, backed by BP Amoco, Clarksons, Royal Dutch/Shell and Cargill, is where ship owners, shipbrokers and cargo owners can conduct business. The new company offers 'a "life-of-the-voyage" solution for all seaborne wet and dry bulk commodity shipping, including freight management services such as market intelligence, chartering, and risk management tools'. Clarksons is the world's largest ship-broking group and its chairman is none other than Gary Weston of Weston, Loblaw, National Grocers, Associated British Foods, and so on. BP Amoco owns or operates an inter-national fleet of 31 crude and product tankers while Royal Dutch/Shell 'moves cargoes on some 140 deep-sea tankers and gas carriers around the world every day'. Cargill's Ocean Trans-portation Business is a global charterer of primarily dry-bulk commodities, including grain and minerals.[124]

- Provision X: In April, Cargill and IBP, along with Smithfield Foods, Tyson Foods, Gold Kist and Farmland Industries, launched Provision X, an online, B2B marketplace for meat and poultry products, service, and information. The marketplace is to be a neutral web-based exchange that will provide a single, convenient place for buyers and sellers of meat and poultry products to connect with each other. The five players, which represent 55 per cent of the beef, pork, and poultry industry in the US, invested a combined $17 million in the project.

- GSX.com was formed by Cargill Steel, Swiss-based Duferco (the largest independent steel trader in the world), Samsung of Korea, and TradeARBED of Luxembourg, which operates a global network of trading operations in steel products.

GSX.com is intended to be an independent electronic exchange for the international trading of steel, as well as offering online financing, risk-management and logistics options as an integral part of the exchange. 'This first truly global Internet steel trading exchange will ... offer a more open and competitive negotiation process so users can have greater control over the costs associated with international steel trading. All individual transactions will be conducted with complete confidentiality between users.'

- Theseam.com: The Seam brings cotton interests together: Cargill's Hohenberg division, Dunavant Enterprises, the Allenberg Cotton Co. division of the Louis Dreyfus Corporation, Plains Cotton Cooperative, Avondale Mills and Parkdale Mills.

- EFS Network, Inc. is an electronic foodservice supply chain network. 'We create solutions to help companies from all segments within the $411 billion foodservice industry to reduce the more than $14 billion in annual foodservice supply chain inefficiencies that have been identified.'

- Pradium.com: Cargill, ADM, Cenex Harvest States, DuPont, and Louis Dreyfus joined forces again to form Pradium Inc. to offer real-time, cash commodity exchanges for grains, oilseeds and agricultural by-products as well as delivering global information resources. Daniel Amstutz, best known as chief negotiator for agriculture in the GATT Uruguay Round trade negotiations during the Reagan administration, is Pradium chairman. Amstutz, of course, was a Cargill executive in the 1960s and '70s.

Pradium reportedly merged with Rooster in February, 2001, but the listed press release came up on my screen as 'this page not available'. Only later, in a story on the death of Rooster, was it reported that Pradium had not merged but collapsed. While searching for further information in other listed press releases, I frequently came upon a notice saying 'access denied'. One has to wonder what was in a press release that Cargill would prefer you to forget about. While searching under 'rooster.com', however, I did come upon this notice:

To our subscribers:
We regret to inform you that effective today [October 12, 2001], Rooster.com will cease operations. In this tough economic climate, we were unable to secure additional funding. We appreciate your support and interest in our information and services. Since launching in the Spring of 2000, we had over 30,000 registered users. Your support helped Rooster.com receive acclaim for its news email, Rooster Call, and as *Forbes* 'Best of the Web: B2B directory' for the past two years.

Sincerely, Rooster.com

At the top of the list of Cargill's 'eVentures' at the close of 2001: www.alterna.com, 'a leading global provider of advanced liquidity management, transaction processing and settlement solutions for corporations, trading communities and their worldwide banks'. In 1997, alterna Technologies Group launched 'The world's first Internet-based liquidity management platform which today forms a unique international e-finance infrastructure connecting corporations and trading partners with the global banking system. alterna has assembled a comprehensive suite of liquidity management solutions ... alterna is a privately-held Canadian corporation based in Calgary, Alberta and with offices in the United States and Europe.'

10   Coming and Going: Transport and Storage

> The development of modern transportation, storage and handling systems has made it possible to move huge quantities of foodstuffs great distances ... Technologically ... it has become possible to depend upon distant food supplies to meet a growing proportion of both basic needs and dietary improvements. – Robbin Johnson, vice-president, public affairs, 1988.[125]

The development of 'modern transportation, storage and handling systems' is a Cargill specialty and a reflection of how the company views the world. Being able to 'source' commodities at will, transport them efficiently to any destination, and deliver them reliably is not only a profitable business in itself, it is also a way of gaining leverage or advantage in the market, whether in bulk or IP commodities.

To source, transport and deliver bulk commodities globally requires a rather special view of the world, a view one can really only adequately get from outer space, from a satellite. Cargill did not, however, start with the satellite view. It started with the economics of water transport and a creative imagination.

A conventional map of North America, for example, displays two coastlines, one on the east and one on the west, and two borders, a long one to the north and a much shorter one to the south. Very few maps reveal the signficance of inland water routes (rivers) and ports, and virtually all distort ecological and geographic reality by imposing political jurisdictions in different colours. W.W. Cargill's successors had the good sense to ignore the political jurisdictions and pay much more attention to water, always by far the cheapest means of transporting bulk commodities.

Since Cargill is most highly developed in North America, we can focus on North America to gain an understanding of Cargill's strategic approach to geography and transportation. Before proceeding with that, however, we must add one other element to the picture. Storage plays an unglamorous but vital role in both the transportation and trading of grains and oilseeds.

Storage

Storage capacity is obviously essential to an effective delivery system. It takes many truckloads to fill one railcar, and many railcars to fill one ship, and transhipping facilities, or terminals, have to have the capacity to store enough truckloads or train loads to fill a unit train or a ship at one go, not over a week or two. Storage capacity, like the reservoir behind a dam, gives the transportation system elasticity and power.

Storage capacity also provides the ability to deliver on demand, or to withhold from market, both of which are essential to successful trading. The amount of leverage a company has on the market, or in negotiations, is in direct relation to its ability to supply at favourable prices or to sit out low prices. This is crucial to the strength of one's hand in playing the futures market. The greater your reserves, the more you can play safely, or at least wittingly. There are overhead charges, of course, but if much of the storage capacity is considered a cost as part of the transportation system, or has been largely paid for by publicly owned port authorities or under one government programme or another, the major cost of storage is that of the grain itself. A company may not use all the storage capacity it has, but it is the capacity itself, not necessarily its maximum usage, that provides the financial leverage in dealing in both real, visible grain and in futures contracts for invisible or even non-existent grain. Grain can materialize, if necessary, before the future contract has to be supplied.

Because it has the storage capacity, it behooves Cargill to take advantage of it for speculative purposes. This may explain why Cargill places so much emphasis on the futures market, and expends so much energy propagandizing farmers to use the futures market to maximize their returns. There is a difference in the strength of the players, however. The individual farmer can play this game only within the very strict limits of the crop he or she actually has to sell. And while the farmers may get a higher return by hedging on the futures market, Cargill will still be using that grain for its own speculative purposes.

While Cargill tries to give the impression that its way of doing business is the only one possible, there are alternatives, among them the pooling of grains and single desk selling characteristic of the AWB and the CWB. These organizations pool individual farmers' grain to accumulate enough to play in the same league as the Cargills and ADMs. The Wheat Boards also have the choice of staying out of the futures market altogether simply by selling all the grain they are responsible for through negotiation of prices and conditions, such as grade, time of delivery, form of payment, etc. Since this deprives the private traders of large quantities of grain for speculative purposes, the companies expend considerable energy trying to destroy these Wheat Boards. The AWB, which was very similar to the CWB, is now half-gone.

It is not just grains and oilseeds that Cargill plays with and moves about; it may be salt, sugar, cottonseed, soybeans, or frozen orange-juice concentrate.

Transportation Systems

Looking at North America as Cargill and the ecologically minded do, one first of all sees that North America really has four coastlines. In addition to the obvious ones on the east and west sides, there is the third, the 3,200 km-long Mississippi River (6,400 km including tributaries) that runs down its middle, right through the corn and grain heartland of the US, and the fourth, the Great Lakes–St Lawrence River complex running eastward from the centre of the continent. (Cargill has actually referred to the St Lawrence Seaway as a 'Fourth Coastline' in its *Bulletin*, and as we shall see, one can view Brazil similarly.)

In the east and the north-west there are also deep-water arteries; the Hudson River in New York state and the 4,800 km Snake and Columbia Rivers system in Oregon and Washington states. All of these waterways lie completely within the territory of the US, except for the St Lawrence River and its seaway, and all except the St Lawrence are maintained at the expense of the US public, which makes them doubly attractive as transportation routes to a company like Cargill. The St Lawrence Seaway relies much more on ever higher toll charges, making it an increasingly costly route – for the shippers, that is.

Maintenance of inland and coastal waterways in the US, including construction and maintenance of locks, dredging, navigational aids and charting, have been the responsibility of the US Army Corps of Engineers since the passing of the Rivers and Harbours Act of 1824. The cost is both substantial and highly volatile because the choice of projects undertaken by the Corps of Engineers is subject to a great deal of political pressure and patronage. Government appropriations

Map 1   Major North American bulk transport waterways

for maintenance of the 18,000 km system of interconnected commercially navigable rivers in eastern and central US were in the billions of dollars annually in the 1990s.*

While Cargill has always been a staunch advocate of the so-called free market and decried government restriction or interference, it has at the same time always been more than willing to have the public bear as much as possible of the costs of the infrastructure it wishes to exploit. For example, in addition to maintenance of the water routes it uses, many of the port facilities leased or owned by Cargill, from piers to terminal elevators, have been built at public expense by local, regional or state agencies. Water and road access to these facilities is also a public expense.

For a company frugal with its capital, but careful to insist on control, a long-term lease can provide the same kind of private control as outright ownership. Where there are no useful public facilities to exploit, as in Santos, Brazil, or India, Cargill is, however, prepared to go so far as to build its own port.

Until the 1930s, Cargill was, as it has said, 'a regional grain merchandiser' with its activities confined primarily to the US interior. Inland water routes such as the Great Lakes, the Mississippi River and the Erie Canal (the New York State Barge Canal, which had been rebuilt and reopened in 1918), were used where possible for domestic distribution.

When the Hudson River was dredged to a depth of 8 m in 1931, all of a sudden, Albany, 230 km up the river from New York City, had the opportunity to become a deep-water port. When Cargill got word that the Albany Port Commission was considering the construction of a terminal elevator, John H. MacMillan hastened to propose that the Commission construct, at a cost of $1.5 million, the world's largest terminal elevator, with a capacity of 13 million bushels (353,800 tonnes). MacMillan promised that Cargill would lease the facility long-term if the Port Commission would copy, on a smaller scale, the company's recently completed elevator at Omaha, including the suspended (cantilever) roof.

The Albany elevator was completed in 1932, leased to Cargill Grain Company, and Albany became the preferred destination for Cargill's grain being shipped east. Grain movement by barge through

---

* For an almost unbelievable chronicle of the politics of water in the US, see Marc Reisner, *Cadillac Desert – The American West and its Disappearing Water*, Viking Penguin, 1997.

the canal to Albany might be considerably slower than by rail, but the net cost was lower. The flexibility and value of the Albany location became even greater when Cargill acquired its own trucks in the mid-1930s and began delivering grain throughout New England, and although the Erie Canal was outgrown and replaced by railroads in the late 1930s, the Port of Albany continued to handle large amounts of grain, particularly corn, from the mid-west for export. The Albany terminal today is capable of receiving 100-car unit trains (see below) from the mid-west.

With the success of the Albany terminal, Cargill began to look more seriously at the potential of the Mississippi River, entering into discussions with authorities in St Louis and Memphis about sites for terminal elevators. Memphis won the bidding for Cargill's business when the Memphis Harbour Commission agreed in 1935 to build a terminal, with Public Works Administration (federal government) funds, according to Cargill's specifications – 'Cargill's aversion to big government notwithstanding', as company historian Broehl puts it.[126]

The next move for Cargill was to begin to acquire its own transportation equipment, in this case barges. Cargill first purchased a few old, small wooden barges for use on the Erie Canal in 1937 and within two years had entered the shipbuilding business for itself, constructing steel barges at its Albany terminal for its own use. It also had towboats and barges built for it in Pittsburgh. During the war years of the early 1940s, Cargill even got into building small ships for the US Navy at what had by then become Port Cargill at Savage, on the south side of Minneapolis. More recently the company has been content to take advantage of the skills and resources of others to build its ships and barges to Cargill specifications and with Cargill innovations. The steel, however, more than likely comes from Cargill's steel-making subsidiary, North Star Steel. The pattern here is strikingly similar to Cargill's other lines of business; Cargill supplies the inputs and buys the product, leaving the riskier segment of the business to independent contractors or growers.

The success of Cargill's waterborne ventures, in design and construction as well as in actual use, has continued with many novel and creative projects in water transport and service. These include its sulphur barge operating out of Tampa and its K-2 in the Mississippi River above Baton Rouge.

Growing competition from overseas steel-makers, however, combined with fewer US orders for steel bars used in building and highway construction have reduced the profitability of Cargill's North Star Steel unit, the second largest US maker of steel bars and the second largest US electric mini-mill steel recycler (which is supplied by another Cargill subsidiary, scrap metal dealer Magnimet), and the seventh largest among all of the nation's steel producers. Signalling a possible exit from steel making altogether, Cargill sold North Star's Tubular Division to Lone Star Technologies Inc. for $430 million in 2001. North Star Tubular produces seamless steel products, including pipes, couplings and casing that are used primarily in the oil and gas exploration and transmission industry. (The steel industry continues to decline. In February 2002 prices were at 20-year lows, imports were up, and the industry was haemorrhaging. 'The state of the industry is unmitigated disaster', said one steel industry consultant.)

Cargill opened offices in Portland and San Francisco and leased a Seattle terminal in 1934 after observing the growing shipment of grain from the West Coast. After World War II Cargill greatly increased its terminal elevator capacity by both increasing capacity at existing locations and having new facilities built, often by local port authorities or cities to Cargill specifications. For example, the Greater Baton Rouge Port Commission built a new 4.8 million bushel terminal to Cargill specifications in 1955 and then leased it to Cargill. It was the first private-company terminal on the Gulf of Mexico (actually upriver about 160 km), giving Cargill a distinct advantage in managing exports.

When the Commodity Credit Corporation (a US Federal agency) was given an increased role in the acquisition and storage of grain with the passing of PL 480 (Food For Peace) in 1954, Cargill was quick to garner the contracts and provide the storage in its publicly built facilities.

Meanwhile, Cargill continued to expand both its barge fleet and its deep water freighters, and by 1992 it was reported that through its international subsidiaries and affiliates, Cargill owned, or had under long-term charter, some 20 ocean freighters, ranking the company among the world's largest dry-cargo vessel operators. It was reported in 2000, however, that Cargill had sold the last four of its Panamax bulk carriers, with an agreement to charter them back for five years. Cargill still retained a strong time-chartered fleet of nearly 150 vessels but said it planned to focus on logistics management. Cargill also owned or operated somewhere between eight and eleven towboats and a fleet of 682 barges. That still left Cargill a relatively small player compared to American Commercial, a subsidiary of CSX

Corporation and the largest barge company in the US with a fleet of 3,500 barges. (It also operates in South America, as we shall see.)

It was not just water transport that was changing the movement and storage of grain. The growing number of trucks on the road encouraged Cargill to think more creatively about the use and location of its smaller grain handling facilities, the country elevators. It was no longer necessary for elevator location to be limited by the distance a horse could travel or how much a team could haul. This led Austen Cargill to propose, in 1940, that the company's country elevator business be reorganized around what he called 'trade centres' and the designation of the elevators in these locations as profit centres. The repercussions of this modest, yet radical change in thinking continue to be felt as elevators are closed and consolidated, ownership becomes more concentrated, and farmers have to bear more and more of the costs of getting their grain to a market facility, whether for domestic use or export.

Cargill continued to expand and integrate its US grain-handling and transportation facilities through the 1950s. Integrated barge-towboat units were built and Cargill's transportation services, including trucking and equipment leasing, were integrated into Cargill Carriers Incorporated. In 1960 the 6,630 horsepower river towboat 'Austen S. Cargill' was launched at St Louis and Cargill's first ever bulk cement cargo was loaded in Antwerp, Belgium.

With the prospect of the St Lawrence Seaway being able to handle ocean-going vessels when it opened in 1958–9, Cargill became concerned that the advantage it had gained with its Albany terminal would be lost. It decided to protect its export position by building a large terminal elevator at Baie Comeau, at the mouth of the St Lawrence River. At 12 million bushels, this was the largest grain elevator in Canada when it opened in 1960. (Albany was 13 million.) It was also the largest single financial commitment ever made by the company (over $13 million – which is peanuts now) and was intended to be the linchpin of a continental grain transport system. Arrangements were worked out so that the terminal could handle Canadian and US grain separately without duty payments.

At the time, Cargill estimated that it had 28.6 per cent of grain exports from the US while Continental handled 25.4 per cent. Dreyfus was estimated to have 17 per cent of the export market and Bunge 10 per cent.

While positioning itself at the mouth of the St Lawrence, Cargill was also developing a receiving port on the other side of the Atlantic.

In 1960 the company acquired terminal facilities in Amsterdam which, like those in Albany and Memphis, were built almost entirely to Cargill specifications by the Amsterdam port authority which then leased them to Cargill.

One of Cargill's most creative approaches to 'port' facilities, and one of the few it seems to have paid for all by itself, is the world's largest floating feed-products transfer facility opened by Cargill at Mile 158 in the Mississippi River in 1982. The $16 million 'K-2' export terminal, run by Cargill subsidiary Rogers Terminal and Shipping Co, was designed to transfer soybean meal, grain and grain products from river-going barges to deep sea freighters at the lowest possible cost, at least to Cargill. The K-2 is a self-contained facility that generates its own power and produces potable water for sanitation and irrigation. The K-2 is capable of continuously weighing and sampling both ingoing and outgoing commodities, as well as blending to customer specifications, at the rate of 1,000–1,200 tons per hour. Cargill has similar floating facilities in Amsterdam which transfer grain directly from ocean vessels to barges, and there may well be others around the world that I do not know about.

Building on its bulk transport experience, and always looking for greater efficiencies in the development of its forms of monoculture, Cargill came up with the idea of the 'unit train' to serve the ever-growing inland elevators. A unit train is a train consisting solely of specially designed hopper cars to be completely filled at one inland terminal and delivered as a unit to either an inland customer or to an export facility, such as Baie Comeau or Albany. Unit trains do create efficiencies in the handling of large amounts of grain, and they do serve the interests of large corporate grain shippers and large inland grain terminals, but they do so at the expense of smaller terminals and customers. With the deregulation of freight rates in Canada and the US, the railroads can offer, or the grain shippers demand, discount rates for loading 75- or 100-unit grain trains at those elevators and terminals that have adequate handling and storage capacity. This puts increasing pressure on the system to eliminate the smaller elevators and forces farmers to travel further to deliver their grain to an elevator. This trend is now causing farmers to come up with alternatives that enable them to regain some control, such as trackside loading of 'producer cars'.

To ensure that its concept was not limited by the availability of suitable railcars, between 1986 and 1992 Cargill commissioned 1,600 railroad tank cars at a cost of $80 million. (One could reasonably assume that a stipulation of the contract for the fabrication of these cars was the use of Cargill steel from North Star.)

Cargill's railroad cars, for corn syrup, grains and other commodities are often, as the company says, 'the only package our customers see',[127] and consequently they have to be clean and well maintained. This goes for their trucks as well. Transport Services Co. in Memphis, for example, is a Cargill subsidiary dedicated to hauling vegetable oils, corn oil from the Cargill wet-milling plant in the Port of Memphis and peanut and soya oils from the recently acquired Kraft plant, on the north side of that city.

With the notable exception of the now abandoned Erie Canal, it is generally the case that low-cost water routes are naturally occurring and cannot be relocated, however much the US Army Corps of Engineers might try. Other means of transport have to be utilized if large quantities of goods and commodities are going to be moved to or from water or between dry land locations. In the twentieth century the railroads filled this need and in spite of the rise in truck transport, the railroads remain second to water for low-cost transport, particularly of the bulk commodities that Cargill deals in. One need only stand in a corn field in central Nebraska within sight of one of the main east–west rail lines to realize how vital the railways are to the continental transportation system. There is a continuous flow in both directions of very long freight trains with one in sight at almost all times.

Along with the centralization of the grain-handling system, with small country elevators disappearing along with the branch rail lines that served them, the rail networks of North America have themselves been transformed by integration and reorientation. All the major railways on the continent have pursued amalgamation and rationalization. 'Rationalization', in agriculture, means the abandonment of small farms in the face of the demands of industrialization. In the case of railroads, it means the abandonment of branch lines – and small towns – in the service of the same ideology. The railways claim that the small branch lines are unprofitable to maintain, but that depends on who pays for the alternative, road trucking. In fact, consolidation and rationalization have been applied to grain elevators, rail lines and farms alike, with uniform results: higher costs incurred by the remaining farms for transportation, higher costs to governments for road maintenance, a smaller tax base, and disappearing rural communities.

11 Typical Stories – Canada and Mexico

Canada

In spite of having established a beachhead in Canada in 1928 as a grain merchandizing company operating out of Vancouver, Cargill remained largely invisible until its 1974 purchase of National Grain. Dick Dawson retired in 1993 after 35 years with Cargill Ltd, the last 19 of them as senior vice-president. When he moved up to this position in 1974, he told me, his first accomplishment was to buy National Grain with the $120 million that he had made trading grain in 1971–2 when the Russians went on a buying spree. According to Dawson, when he asked Whitney MacMillan what to do with the money, MacMillan told him to 'buy something'. What he bought was 286 country elevators, five feed mills, a terminal elevator at the head of the Great Lakes at Thunder Bay, and a significant 'originating' capacity in one of the world's major grain-growing regions.

On the occasion of Dawson's retirement, reporter Allan Dawson (no relation) wrote in the *Manitoba Co-operator*, 'A good portion of Dawson's job with Cargill involved exercising influence. And some would say he did his job well ... In fact, some might speculate Dawson is retiring because the policies he has promoted for years are finally being implemented.' There was another word going around Winnipeg, however: one of the reasons Dawson retired early was that he was just a little too gregarious for the liking of the corporation. This sounds probable to me since we had several conversations and he once spent a whole morning with me discussing the company's business. That was before it was decided, as I was told by another senior executive, that: 'It is not a good use of our time to talk with you.'

In spite of the significant physical presence achieved with the purchase of National Grain, Cargill's position in the trading of wheat and barley for export remained that of every other grain company and cooperative, that is, acting as agents of the Canadian Wheat Board. It is not surprising then that Cargill has laboured long and hard, with much deviousness and many front groups such as the Western Canadian Wheat Growers Association, to marginalize or destroy the Canadian Wheat Board. But even if the Wheat Board could be destroyed, there were still the Prairie Pools to contend with. Cargill would have to acquire control over, if not own outright, sufficient modern infrastructure to provide the foundations of an alternative grain handling system. So it replaced many of the old grain elevators it had acquired from National Grain with a few high capacity elevators, such as the inland terminal at Rosetown, Saskatchewan, opened in 1976, but true to form, it sought ways of gaining effective control of the facilities it needed without having to invest its own capital in them. When Canada's first producer-owned inland terminal* was built at Weyburn, Saskatchewan, Cargill was designated as sole selling agent for the terminal's grain. Not only does such an arrangement avoid the commitment of large amounts of capital, it also gives the company a great deal of strategic flexibility.

The success of the Weyburn terminal, combined with the hostility of the private traders toward the Canadian Wheat Board, convinced farmers in the north-eastern area of Saskatchewan's grain land that they, too, should build a big inland terminal. North East Terminal Ltd opened in 1992 and though it is supposedly an independent elevator, Cargill Ltd holds 25 per cent of the company in return for an investment of $500,000 and has a contract to operate the terminal. While the farmers who initiated the project raised $1.8 million in their initial share offerings, because their 75 per cent share is split up among many farmers, Cargill has, whether the farmers realize it or not, effective management as well as operating control.

By 1982 Cargill had become the leading private exporting agent for the CWB, handling 8 per cent of all prairie grain, but for all its efforts to destroy the Wheat Board and to build an alternative infra-structure to the Prairie Pools, by 1994 Cargill had only managed to increase this to 10 per cent.

Cargill entered the Ontario grain market, which is not within CWB jurisdiction, in 1978, establishing a trading office in London,

---

* Historically, a terminal elevator has been an elevator located at the end of a rail line at a seaport where grain was cleaned and stored before being loaded on a ship. The term 'inland terminal' is now used to describe an elevator located inland, on a main railway line, where grain can be cleaned to export standards and stored before being loaded onto unit trains for direct shipment to a seaport for export.

Ontario and buying Erlin Grain in Talbotville, Ontario, to give it a grain originating capacity.

Cargill's expansion in the feed business in Canada paralleled that in the US. When it purchased National Grain in 1974, the package included five feed mills. In 1985 or 1986 it purchased Kola Feeds in Brandon, Manitoba, and then Southern Feeds in Lethbridge, Alberta, where it already had a molasses-based liquid feed supplement plant. This made Cargill the top feed supplier in the major cattle-producing region of the country, but it did not stop there. Cargill expanded its fertilizer services by adding four small blending facilities, starting eleven others, and buying two fertilizer sales operations in Saskatchewan and Manitoba.

Cargill continued its relentless expansion in the retail feeds business, establishing a beachhead in Ontario in 1987 with the purchase of Ayr Feeds, primarily a supplier of feed to the poultry business. At the time, Cargill owned nearby Shaver Poultry, a major source of laying hen 'genetics', otherwise known as breeding stock. The next year it gained a lot of new territory with the purchase of the feed mills of Maple Leaf Mills from Hillsdown Holdings for $40 million. (Hillsdown had bought Maple Leaf Mills from Canadian Pacific the year before for $361 million.) The deal gave Cargill, at a bargain price, 23 country elevators in south-western Ontario and four grain terminals: Midland, Port McNichol and Sarnia, Ontario, and Saint John, New Brunswick. The three Ontario terminals were situated to receive grain by water from the west via the Great Lakes and load railcars for further shipment to east coast deep-water ports in the era before completion of the St Lawrence Seaway. The St John, New Brunswick, terminal was a deep-water terminal at the end of the rail line. The acquisition of Maple Leaf Mills also made Cargill Canada's largest soybean handler.

Cargill's next move in Ontario was to purchase Arkona Feed Mills in Arkona, Ontario, in 1989. Cargill spent $1.5 million upgrading the plant so that it could serve the specialized pig and dairy industries of Michigan and Ontario with its Nutrena feeds. The Arkona management now reports to Cargill's feed plant in Mentone, Indiana. This transnational integration was not a fruit of free trade agreements, since feed grains already moved freely across the US–Canadian border, but simply one expression of Cargill's long-term continental strategy and its 'ecological' perspective. South-western Ontario, southern Michigan and northern Ohio and Indiana constitute a kind of bioregion.

In 1989 Cargill also bought the fertilizer operations of what was then Cyanamid Canada, giving Cargill a significant 'retail presence' in 22 locations in Ontario and Quebec. It immediately replaced two of the older facilities with a new $2 million plant in Harrow. Cargill's manner, or lack of manners, in such transactions was reflected in the experience of Cyanamid employees at the time of Cargill's takeover. When I visited the Cyanamid fertilizer outlet near Alliston, the woman in the office told me that while nothing had changed, the only information they had about the deal was what they read in the newspaper. The assistant manager in another Cyanamid facility said that when Cargill bought Cyanamid's fertilizer business nothing was said to the staff and they were not even asked if they wanted to work for Cargill.

Today Cargill operates 24 Farm Service Centres in Ontario and is a joint-venture partner in ten other Ontario grain and crop input businesses.

Cargill moved into actual fertilizer production in Canada in 1989 with the construction of one of the world's largest nitrogen fertilizer plants in Belle Plaine, Saskatchewan, where it could obtain the necessary natural gas feedstock from a transcontinental pipeline (see Chapter 12).

Then in 1991 Cargill purchased Alberta Terminals Ltd (ATL) from the Alberta government for the bargain price of $6 million. ATL consists of inland terminals at Lethbridge, Calgary and Edmonton and an off-track loading facility in the Peace River region of Alberta. The facilities had cost the Alberta government $17.9 million since it acquired them in 1979. Cargill said the deal, which gives it more than half of Alberta's primary elevator storage capacity, put it in a good position to ship grain into the US.

Recent events at Cargill's Lethbridge facility illustrate not only the company's willingness to accept public subsidies, but to actively pursue them to the point of what can only be called blackmail. For example, Cargill laid off its eight employees there in 2001 and said it might close the terminal altogether. Two months later Cargill announced the facility would remain open after it had worked out a plan whereby Employment Canada would pay the facility's six unionized workers for two of the three days they work to keep the plant operating.

By the end of 1996 Cargill Ltd (Canada) owned 75 primary and terminal elevators, employed 3,400 people and reported sales for fiscal 1996 of C$3.2 billion of the company's global revenues of $56 billion, the company record.

Mexico

Cargill has had a rather contrasting presence in its very different neighbour to the south.

Cargill established its beachhead in Mexico in 1964 with the opening of a molasses trading office in Mexico City called Carmela. Three years later it established the grain trading and agricultural brokerages Carmex and Carmay. In 1971 Cargill acquired C.C. Tennant Sons & Co. and began trading minerals under the name Tennant Mexico. When Cargill acquired Hohenberg Bros Co. in 1974 it became a cotton broker in Mexico as Empresas Hohenberg, Industria Hohenberg, and Empresas Algodonera Mexicana.

Heinz Hutter, then president of Cargill, offered his recommendations on how Mexico could improve its agricultural production in a talk at the University of Illinois in 1990. Among his recommendations: increase farm size; convey land titles to communal farmers; make it possible for 'efficient farmers' to buy land freely.* Two years later, in mid-1992, the Mexican government amended the country's constitution (Article 27) to enable it to implement the policies recommended by Cargill, among others, and began allowing mills to apply for import licenses for wheat. Previously, all wheat imports were coordinated by CONASUPO, the government food distribution agency that had nearly complete control of the importing and internal marketing of wheat and corn.

The Mexico City newspaper *El Financiero* carried a substantial article on Cargill in Mexico in 1993. Describing Cargill as 'already a major supplier of corn to Mexico', the article commented: 'Historically, the company has tended to dominate the markets it enters. Its presence in other countries raises questions over who will control

---

* One of the most important changes in the constitution pertained to the *ejidos*, the communal farms that gave Mexican peasants security of tenure on the land they worked. This was one of the major achievements of the Mexican revolution, but it was also a major obstacle to the 'modernization' or 'rationalization' of Mexican agriculture. The *ejidos* prevented farm consolidation and industrialization. What Cargill and others were calling for was the removal of this obstacle to their freedom. The constitutional change that allowed peasants to gain title to the land simply cleared the way for the land to be acquired, one way or another, by others.

Mexico's food supplies.' The article quoted Cargill spokesman Greg Lawser as saying: 'The company's operating philosophy throughout the world is to begin with reasonably small capital investments in fields where we believe we can bring expertise and technology of use to the area, and to grow our business from that small beginning.'

El Financiero pointed out that this corporate strategy conveniently dovetailed with the development policy of President Salinas de Gortari to encourage big investors to take over small enterprises, especially in agriculture. Commenting that 'information about Cargill's designs on Mexico is difficult to obtain', *El Financiero* provided Lawser's response: 'The commodity business doesn't talk about its strategies.'[128]

12   Fertilizer

Fertilizer quite naturally keeps cropping up in the Cargill story. It is, after all, used in great quantities in the industrial production of crops, and if one is in the bulk commodity business, there are not many more bulky commodities. If one is selling seed and buying grain, then selling fertilizer to the same people you are already dealing with is just common sense. Cargill has not really used fertilizer to establish its beachheads, but as we have seen it is apt to follow close behind.

The three major components of commercial fertilizer are nitrogen (N), phosphorous (P) and potassium (K). The nitrogen is produced from natural gas, the phosphorous comes from rock phosphate, and the potassium comes from potash. The latter two are mined, and Cargill is involved primarily in the first two.

Phosphate

Most of the world's phosphate rock is mined for processing into phosphate fertilizer. Morocco in north Africa contains at least half of the world's phosphate rock reserves and is the world's largest unprocessed-phosphate-rock exporter, though in recent years it has built up a considerable capacity for conversion of rock into fertilizer and chemicals.

The conversion of phosphate rock into a soluble form for use as fertilizer requires sulphur and large amounts of ammonia to produce its most common form, diammonium phosphate (18–46–0) generally known as DAP. The US produces 60–65 per cent of the total global production of 24 million tons of DAP. US domestic sales of DAP in 1990 amounted to $1 billion out of a total of $7 billion spent on all fertilizers. Total global trade in DAP is about 14 million tons per year, with the US accounting for 65 per cent of this and Morocco 15 per cent.

One quarter of the world's phosphate is to be found in central Florida under 1.5–15 m of sandy soil in two locations, referred to as the Northern District and Bone Valley. The two sites provide about 80 per cent of US phosphate production. The 100 km^2 Bone Valley deposit was discovered in 1881 by the US Army Corps of Engineers, but it was created something like 15 million years earlier when seas covered the area and deposited the remains of billions of tiny sea organisms in the beds of sand and clay. To get at the phosphate rock, the overburden is first scraped aside and piled for future use in site restoration; the actual phosphate ore is removed by means of giant draglines. It is then washed, spun, crushed and vibrated, yielding fine clay suspended in water, coarse phosphate rock, and a mixture of sand and fine phosphate which is then processed further. The slurry is piped to settling ponds, the sand is stockpiled for future mine restoration, and the phosphate is hauled by rail and truck to the processing plants. Because phosphate rock is not water-soluble, treatment with sulphuric acid is essential to convert the rock into fertilizer and other products.

When Cargill purchased the Gardinier fertilizer plant on Tampa Bay in 1985 from the Gardinier family of France, it had already been mining phosphate at Ft Meade in Bone Valley and shipping the ore to the Gardinier fertilizer plant for proccessing. The plant, built in 1924 by the US Export Chemical Company with the most advanced technology of the time, was in bankruptcy proceedings at the time. One of Cargill's first acts was to lower wages. As one employee put it: 'Cargill made us pay a bit for the generosity of the Gardinier family, which treated us very well – perhaps too well.'

The plant was run down, and in 1988 154,000 litres of phosphoric acid spilled, killing the fish in the Alafia River. Cargill was fined $2.2 million and ordered to upgrade the facility.[129] Cargill, in a characteristic manner, paid the bills, did the clean-up and upgrading at a cost, its says, of $125 million, and has since gained public recognition for its environmental conscientiousness.

From the 3.5 million tons of phosphate rock it mines each year in Bone Valley, Cargill produces 7 per cent of the total US phosphate fertilizer supply, primarily in the form of DAP. Only 15–20 per cent of the plant's production goes to the US or Canada and the 85 per cent that is exported leaves the Port of Tampa by water.

Cargill figures it has 15 years of phosphate ore reserves on hand underground at all times, while a constant swapping of land goes on so that companies can mine contiguous parcels. Some land is owned, some leased. The full cycle from removal of overburden and mining through complete reclamation is three years, as required now by state law.

Sulphur, as has been noted, in the form of sulphuric acid is a key ingredient in the production of phosphate fertilizers. Cargill uses about 600,000 tonnes of sulphur annually, which it brings in molten form from Mexico, Texas, Louisiana and Carribean ports. Drawing on its skill and experience in bulk transport by water, the company designed a special barge (S/B)/tug unit and had it custom-built for this purpose. The 130-m S/B Alafia was christened in January, 1991.

Phosphate mining, like all open-pit mining, leaves horrendous scars on the earth. In recent years public pressure has led to legislation requiring the restoration of mine sites. In Florida, as mentioned above, companies are now allowed three years from commencement of mining to full restoration of the site.

Still, what one sees while travelling the back roads of central Florida is very depressing. Bone Valley, south of Highway 60 east out of Tampa, in the centre of the state, is a moonscape of overburden piles and gaping scars. In the midst of this forlorn landscape, however, as I tracked down Cargill's Ft Meade mine site, I came upon the lands the company had restored. I was not expecting to see citrus groves and dryland cattle feeding on verdant pastures on restored mine sites. Discussing this with the mine manager on a Sunday morning when no one else was around, I learned – and I could see it with my own eyes – that Cargill is experimenting with various restoration schemes, dependent on the character of the overburden being restored. It looked to me as if it was being genuinely creative.

Cargill's strategy is to match potential uses of the land with the water and overburden available. Some land is being turned into wetlands, some into citrus groves (800 hectares), some into blueberries (15–20 hectares I was told), and some into alfalfa if, for example, it has more clay that can benefit from the long tap roots of the alfalfa. I saw a beautiful herd of mixed cattle on some of this alfalfa land. In 1994, for the first time, the oranges harvested from Cargill's 800 hectares of groves on reclaimed land will be processed by Cargill's juice plant, which is very nearby at Frostproof.

In May, typically taking advantage of a severe depression in fertilizer prices, Cargill moved into the position of second largest phosphate fertilizer producer in the world with the $150 million purchase of Seminole Fertilizer from Tosco Corporation. The plant had earlier been owned by W.R. Grace & Co. The Seminole operations produce 750,000 tons of phosphoric acid per year, while Cargill produces 830,000 tons at its Tampa plant. The acquisition also gave Cargill an additional mine in Ft Meade and another at Hookers Prairie along with fertilizer plants in Riverview and Bartow, right on Highway 30. Being inland plants, they have to be served by truck and rail transport. Cargill reported that nearly 100 railcars of sulphur are required each month at the Bartow plant, while 1,800 railcars leave the plant with fertilizer every month. Cargill's total phosphate production gave it about 14 per cent of the US market.

When I was looking for the Cargill mine at Ft Meade, I stopped to ask directions from a telephone line crew. They gave me directions all right, but also volunteered that when Cargill bought the Seminole plant they made all the employees sign resignation slips and submit their resumé. They then hired back those they wanted, at lower wages and with only two weeks vacation, no matter how much they had accumulated, which for some long-term employees was as much as five weeks. 'What they want is control, so they can make more money', is the way one of the men put it. 'They want to drive everyone else out, just like Agrico and IMC!' Then they wanted *me* to tell *them* who Cargill really is!

IMC-Agrico and Vigoro Corporation are other partners in the phosphate oligopoly. Prior to 1992, IMC Fertilizer Inc. owned 39 per cent of the US phosphate rock capacity and produced 15 per cent of its phosphate fertilizers, while the Agrico division of Freeport-McMoRan Resource Partners of New Orleans was the second largest phosphate fertilizer producer. Since then, IMC and Agrico have merged their US phosphate operations into IMC-Agrico.

In 1999, IMC Global Inc., CF Industries Inc. and Cargill Fertilizer Inc. formed a joint venture to remelt sulphur for use at their respective Florida phosphate fertilizer operations in the production of phosphate fertilizers. The facility, Big Bend Transfer Co. is located in Tampa. CF Industries, Cargill Fertilizer and IMC Big Bend Inc., a wholly owned subsidiary of IMC Global, will each have equal ownership in the joint venture.

Lithuania seems to be the only country, outside of China and North America, in which Cargill has actual fertilizer production as distinct from fertilizer blending, which uses already manufactured components. In 1999 Cargill acquired a 15 per cent stake in Lifosa, a fertilizer group in Lithuania which produces phosphoric acid and DAP fertilizer.[130] 'The excellent quality of fertilizers made by Lifosa will enable Cargill to further enhance products and services offered to customers in western Europe, and to develop greater knowledge about phosphate fertilizer needs in eastern Europe and nations of the Commonwealth of Independent States', said Henk Mathot, president of Cargill's worldwide fertilizer operations. He noted that Cargill has been purchasing and distributing approximately 50 per cent of Lifosa's phosphate production since 1996. Lifosa is developing a port operation in Klaipeda, Lithuania, and is a 49 per cent joint-venture partner in a compound fertilizer plant adjacent to its phosphate facilities in Kedainiai.

Nitrogen

A few years earlier, in 1989, Cargill began construction of one of the world's largest nitrogen fertilizer plants, in partnership with the provincial government, at Belle Plaine, Saskatchewan, Canada. The province and Cargill formed Saferco Products Inc.* to build and own the $435 million plant, with Cargill holding 50 per cent of the company, the Saskatchewan government 49 per cent, and Citibank Canada the remaining 1 per cent, thus giving Cargill effective control. Cargill also got exclusive marketing rights to the plant's production. Cargill voice Barbara Isman told me that Cargill went for a joint venture at Belle Plaine because then the public support could not be defined as a subsidy in the context of the Canada–US trade agreement.

In announcing its project, Cargill said that the Belle Plaine plant would supply the prairies, Ontario and Quebec. There is no way, however, that Canadian farmers could ever use the amount of nitrogen fertilizer the proposed plant would produce: 125,000 tonnes annually of anhydrous ammonia and 660,000 tonnes of granular urea. On the other hand, the Mississippi River is closer than Vancouver (and it's 'downhill' with no mountains), making it feasible for the plant to be a world-class operation, as Isman described it to me. Belle Plaine is just west of Regina, and just west of Belle Plaine a mainline railway branches off in a south-easterly direction through Weyburn and Estevan, Saskatchewan, and on down to Minneapolis–St Paul at the head of the Mississippi River, where Cargill Port is situated. A bonus for Cargill is that urea is also a source of protein for livestock feed and could be well utilized by Cargill's Nutrena feed division.

More fundamental to the project than transportation was the availability of natural gas, the feedstock, which is readily available at Belle Plaine from the pipeline that runs by the door. The supplier is SaskPower, a provincially owned Crown Corporation. The conservative provincial government had intended to privatize SaskPower,

---

* In 1991 Saferco Products Inc. changed its name to Saskferco Products Inc.

but gave up such plans late in 1989 due to overwhelming public opposition, in part based on public concern over unlimited sales of gas and other natural resources to the US. A fertilizer plant, however, is an efficient way for a private corporation to export natural gas in the form of liquid and granular fertilizer.

While claiming that the project would not require any public subsidies, the provincial government agreed to provide a loan guarantee for the $305 million required for construction, with the remaining $130 million to come from equity shareholders. Thus for $65 million – and only $65 million worth of risk exposure – Cargill got a $435 million fertilizer plant, with the public holding the remainder and the risk. Cargill also obtained the right of first refusal if the province ever decided to sell any portion of its share.

13 The West Coast

With the advent of air travel, seaports have lost the significance they once had for most people. Port facilities are now built on the wastelands of New Jersey or on artificial land made inaccessible to the public for security reasons. Like the K-2 in the Mississippi River, out of sight over the levee, the large grain terminals are now built where few people will ever see them. While one can easily observe the loading and departure of a 75-car unit train of corn from Central City Nebraska, one is unlikely to observe its unloading in Pasco, Washington, and the loading of the grain onto a barge for a trip down the Snake and Columbia Rivers to Portland where it is stored and eventually loaded into a freighter headed for Japan.

The northernmost grain handling port on the west coast of North America is 1,500 km north of Vancouver at Prince Rupert, British Columbia. Opened in 1985, the 210,000-tonne grain terminal was a consortium of six companies widely differing in business philosophy: the three Prairie Pools, United Grain Growers, Pioneer Grain and Cargill. Initially, each company put up $55 million and the Alberta government financed the remaining $220 million, including a $106.25 million mortgage at 11 per cent. Under the agreement, any cash earned goes first to paying operating expenses, then to the participating shippers, and finally to pay off the mortgage. It was not until 1991 that the port made enough to even start paying the interest on the mortgage, which until then had been accumulating and added to the mortgage principal. Its future is as uncertain as that of the rest of the massive infrastructure built in the last decades of the last century when the 'experts' were sure the global trade in grain would expand forever. Cargill was not one of those 'experts' and wisely let others tie up their capital in immobile concrete.

The port of Vancouver, in the south-west corner of Canada, is a good example of Cargill's ability to conserve its own capital and make use of the investments others have made. For years Cargill used other companies' facilities, even though the city is the primary port for Canadian grain exports. Cargill has loaded and shipped containers for export, such as canola and malting barley for China, through Columbia Containers, utilizes the Cascadia Terminal of which it shared ownership with United Grain Growers (which is now

Agricore United) and utilized the largest terminal elevator in the Vancouver port area, which belonged to the Alberta Wheat Pool (which is now also part of Agricore United) for virtually all of its grain exports.

Quite a stir in an apparently calm sea was caused by the announcement on Christmas eve, 1995, that Cargill and Saskatchewan Wheat Pool (SWP) had formed a joint partnership to develop a giant grain terminal at the Port of Vancouver's Roberts Bank facility, which is a very large jetty on the south side of Vancouver well away from the city's congested inner harbour. The jetty was already the site of a large coal-exporting facility. Vancouver Port Corporation said it awarded the right to build the terminal at Roberts Bank to 'SWP/Cargill' because of the companies' ability to 'source product', their financial capability and their proven development experience. The terminal was to be served by CP, CN and Burlington Northern Railways and would handle fertilizer and other products in addition to grains and oilseeds. At the time, Saskatchewan Wheat Pool was Canada's largest agricultural cooperative, handling 30 per cent of all grain deliveries in the prairies, while Cargill had been unable to get its share of the grain market above 11 per cent. The proposal was kicked around between various regulatory agencies for some time, then late in 1997 Alberta Wheat Pool announced the sale of a half interest in its Vancouver grain terminal to Cargill. Alberta Pool said it had no choice since Alberta Pool had handled all Cargill grain shipped through Vancouver in recent years and losing that grain to Roberts Bank would have seriously cut into the terminal's profits. A few months later (March 1998) the Roberts Bank proposal died when the joint venturers' lease option on the project site expired. The Vancouver Port Corp had 'declined to extend' the partners' option to lease the site. According to SWP, the real reason was Cargill's unwillingness to proceed with the project. Cargill was simply being realistic in the face of declining grain exports.

Not very far south of Vancouver are the Ports of Seattle and Tacoma, Washington. Cargill has operated a major terminal at Pier 86 since 1970 when the Port of Seattle completed construction of a $17 million, 17-hectare terminal and leased it to Cargill. This was typical of Cargill's acquisition of port facilities for the past century: the public, through bond issues and/or one public agency or another, pays for the construction of the terminal to Cargill specification with an agreement that Cargill will lease it. Cargill gets the business and the public gets the bill – and some of the business resulting from the presence of the terminal or port facility in their community. In the case of Pier 86, the lease was a bit unusual in that Cargill paid a customary fixed rental but collected dockage fees from vessels using the terminal, remitting only 50 per cent of the fees to the Port of Seattle.

In 1998 Cargill, deciding it needed a better deal, told the Port of Seattle it was thinking about moving and asked the port to forgo its share of dockage fees for six months. In a time of declining grain exports, there were numerous other potential locations for Cargill, should Seattle not want to agree to its demands, including the 3 million bushel grain terminal operated by the Port of Tacoma only a short distance to the south up Puget Sound. The Port of Seattle commissioners agreed to Cargill's demand on the condition that Cargill renew its lease for at least five years.

Soon afterward, Cargill acquired the worldwide commodity marketing business of Continental Grain Company, the fifth largest private corporation in the US with current revenues of $16 billion. The transaction included Continental's grain storage, transportation, export and trading operations in North America, Europe, Latin America and Asia, but not its domestic and international poultry, pork, cattle, aquaculture, flour-milling, animal-feed and nutrition businesses, nor its liquefied petroleum gas trading and financial services businesses. Terms were not disclosed, but the deal was valued by private analysts at about $1 billion. Continental, which is changing its name to ContiGroup Companies Inc., said the deal would enable the company 'to concentrate our financial and managerial resources on significant opportunities worldwide in our fast-growing, higher added-value agri-industries, financial services and private investment operations'. (Paul Fribourg is chairman and CEO of Continental.)

Cargill acknowledged that it and Continental together handled about 35 per cent of US grain exports, but the US Department of Agriculture estimated that the two companies shipped about 42 per cent of all US corn exports, 31 per cent of soybeans and 18 per cent of wheat. The two companies could actually control a much larger share of total US exports because of intra-company exports which are not monitored by USDA, the department said.

To complete the sale to Cargill, Continental was ordered to divest itself of seven US elevators and terminals. It simply decided not to renew its minority interest lease on two, three were sold to the Louis Dreyfus Corp and two to smaller regional companies. When the dust settled, five of the ten facilities ordered divested had become part of Louis Dreyfus, which had revenues of $18 billion in 1999. Its acquisition of the divested facilities raises an interesting question as to its relationship with the other major US grain traders. In 1993 Dreyfus and ADM established a joint venture under which ADM assumed operational control of most of the US grain elevators owned by Dreyfus. But that's a story for another book.

As a condition of its approval, the US Justice Department also required Cargill to get rid of nine grain handling and transport facilities, including its leased facilities in Seattle and Sacramento. Having been ordered to divest its hold on the 4.2 million bushel terminal in Seattle, Cargill simply took Continental's position in the TEMCO elevator in Tacoma, leaving the Port of Seattle holding a worthless $14 million facility.

The largest grain export facility on the US west coast lies a little further south at Portland, Oregon. There Cargill leases one of the Port of Portland's two terminal elevators. Portland is at the mouth of the Columbia River, which flows south from its origins high in the Canadian Rocky Mountains (with great impediments – dams – at frequent intervals along its rather altered route) into Washington state where it is joined by the Snake River at Pasco and then turns abruptly westward. The 750-km Columbia–Snake River system is second only to the Mississippi in the US and substantially reduces the cost of moving grain west to export position by shortening the rail journey. Cargill ships grain by barge from its terminal in Pasco down the Columbia River to its export terminal in Portland. Barge traffic also moves up the Snake River all the way to Lewiston, Idaho.

The third major port area is San Francisco Bay, where the ports of Sacramento and Stockton actually lie about 120 km inland by deep-water channel from the San Francisco Bay bridge. Being that far inland puts them in the middle of the central valley, California's irrigated, fertilized and toxified 'Garden of Eden'.

In the Port of Sacramento Cargill had only a small leased elevator which it gave up, after 36 years of operation, as part of the deal to buy Continental, but at Stockton one can find a complete array of Cargill's integrated business activities: a Nutrena feed mill; a new flour mill located next to the feed mill at the south end of the city; a feed-supplement plant located in the port that utilizes molasses from Hawaii to produce 'Mol Mix' liquid feed supplement; a food-grade corn syrup distribution terminal to receive corn syrup in bulk in Cargill's own railcars and distribute HFCS, sweetener blends and corn-derived products to food processors of all kinds in both Canada and the US; a fertilizer plant (Co-pal Fertilizer) purchased in about 1990; and a terminal elevator to handle the import and export of grains. Cargill no longer owns the molasses business, having sold its global molasses liquid product division to ED&F Man Group for $48.5 million in 1997.

While visiting the Cargill facility at Stockton I noticed that the weathered name on the silos of the Nutrena feed mill was 'Kerr Gifford', the name of the company Cargill had bought in 1953 when it first expanded to the west coast. Cargill also has extensive facilities around the mouth of the Mississippi of course, from New Orleans to Galveston, Texas, in addition to the already mentioned K-2 facility upstream from New Orleans.

On the other side of the world, Cargill appears to be initiating the practice of port development it has so long practised in North and South America. In 1999 it opened a new grain storage terminal at Rota, on the Turkish gulf of Izmir, in 1999. Rota is the most efficient port terminal in Turkey, designed primarily for bulk grain and oilseed imports. 'Turkey is self-sufficient in wheat, but the problem has been quality', explained Cargill's merchandizing manager in Turkey. 'Turkish people like nice white and spongy loaves of bread. For that, you need high gluten, high protein wheat.' Rota will help Cargill import quality wheat in panamax vessels, store it and discharge it to unit trains for delivery to flour mills. The facility also will import corn to serve Cargill's corn-milling operations during times when Turkish corn is unavailable. Cargill bought the Vanikoy corn mill in 1989, and it is building a modern new fructose plant at Ornhangazi. Cargill employs around 450 people in Turkey and in addition to its two grain-processing plants, it has a hazelnut-processing plant.

14   Rivers of Soy – South America

Before reading this chapter, find the best map of South America you can and try to identify the major rivers. The map on page 122 is provided as a starter for you.

Cargill has had operations all over the South American continent for many decades. Some of these provided services and inputs to agricultural production (seeds and fertilizer), but most of its investment is in the gathering and exporting of major crops such as sugar and soybeans. It's only natural then that Cargill can be found on the waterways of South America as well as those of North America and Europe. And for the same reasons: water is the cheapest way to transport bulk commodities and it is common practice for waterways to be maintained largely at public expense. This may be partially explained by the fact that historically rivers have served as boundaries of political jurisdictions, thus falling into a kind of shared/no responsibility situation. Or it may simply be that it was better to have the cost of maintenance born by the state than to have feudal barons demand tolls for passage through 'their' water. The Rhine in Germany is an obvious example, and the possibilities for toll collection can be observed by any river traveller. Today freight traffic on the Danube, which comprises a 2,500-km system administered by The Danube Commission, is free except for pilot fees on the lower 270 km of the river. Costs of maintaining the river, including the locks, are borne by the bordering (riparian) states.

But Cargill's presence in South America did not start on the waterfront. W.G. Broehl records that in 1947 Cargill entered into a business arrangement with Nelson Rockefeller in Brazil, where Rockefeller wanted 'to demonstrate that private capital organized as a for-profit enterprise could also upgrade the economics of less-developed countries'.[131] To do this, Rockefeller had formed the International Basic Economy Corporation (IBEC) in Brazil as a family-held business. Among the enterprises intended for this business were a hybrid-seed-corn company, a pig-production company, a helicopter crop-dusting company and a contract-farm-machinery company. This was, of course, even before all the

Map 2   Major rivers of South America

Rockefeller Green Revolution initiatives of the 1960s and 1970s. In 1948, Cargill Agricola e Comercial was established in Brazil as a joint venture with Nelson Rockefeller's IBEC. In its company literature, however, Cargill identifies 1965, when it invested $9 million in a hybrid-seed-breeding programme and plant, as the beginning of its presence in Brazil.

Though Cargill says it reinvests everything it makes in a country, there is no way of knowing whether this is true or not, since there are no public financial statements. On the other hand, why send the money home? Mother does not need it. One company source stated that from 1967 through 1975 Cargill invested $9.3 million in Brazil, while taking out only $773,000 in dividends, leading to capital formation inside Brazil of $87.8 million by 1975. The only figures Cargill provides now are $600 million in physical assets and annual revenues of about $2.3 billion.[132] Given the emphasis on Cargill's Brazilian website, Portuguese version only, on how much they have continued to invest in Brazil, it is quite plausible that they do reinvest everything they make in the country. Cargill Brazil is second in size only to Cargill's US operations.

One also needs to remember that not having to satisfy the greedy demands of public shareholders, Cargill enjoys a terrific advantage in flexibility and leverage in its capital management over more conventional capitalist enterprises. This is surely what makes it possible now for Cargill to enter so many joint ventures while still being certain that with a 51–56 per cent share they have secure control.

In 1996, Cargill Agricola claimed to be Brazil's largest agricultural company, with 20 production plants and 59 other locations in the country and 4,500 employees. In 2000 Cargill said it was Brazil's largest soybean and sugar exporter, and among the top soybean processors and citrus businesses, with 4,000 employees in more than 70 locations. (Explore Cargill's network of websites, in English and Portuguese, through www.cargill.com.)

With hybrid seeds went fertilizer, and as Cargill puts it, it has two plants, 'distributors in many locations', and recently acquired controlling interest in two other fertilizer companies, Solorrico and Fertiza. The next Cargill project was soybean processing, starting in 1973 with a plant in the southern state of Paraná. The second came along shortly (1975) in Mairinque, São Paulo state. The next three plants are in Uberlandia, Minas Gerais (1994), Trés Lagoas, Mato Grosso (1996) and finally Barreiras, Bahia (1988). The locations of its plants directly reflects the spread of soybean growing in the country. Connected with soybean processing is refining, and Cargill has pursued a somewhat unique, for Cargill, strategy of marketing under its own brand names a wide variety of vegetable oils (soy, corn, canola, sunflower) for the retail market.

While Cargill's Uberlandia started as a soybean-processing plant, it has expanded with corn to become the largest Cargill plant outside the US. It also processes sorghum and in 2000 Cargill opened a citric acid plant in Uberlandia that can utilize cane sugar as well as corn as the fermentation feedstock. Cargill said it was cheaper to use cane sugar than corn sugar (dextrose). The new plant made Cargill the world's third largest supplier of citric acid, which is used to flavour and preserve sodas, fruit juices and dairy products, as well as in the production of medications, cosmetics, plastics and biodegradable detergents. Besides citric acid, the plant's products include potassium citrate, sodium citrate and liquid citric. (In the 1920s, scientists discovered that the spores of a microorganism, Aspergillus niger, can convert sugars into citric acid through fermentation and several years ago Cargill came up with a novel liquid-extraction process for separating out the citric acid from the fermentation.)

Other recent additions to its processing enterprises are a wheat-flour mill in São Paulo state and a cassava starch plant in Paraná state acquired from Grupo Maggi in 2000.

Already established as a cocoa trader, Cargill built its own plant for processing cocoa at Ilheus, Bahia, in 1980. The enterprise was not a great success and the output was of poor quality until 1986 when Cargill acquired Gerkens of the Netherlands, one of the world's largest cocoa processors, and put Gerkens' people in charge of the plant. It now produces top quality cocoa for the world market under the Gerkens name. (In the US, food producers and processors are encouraged to use the products of Cargill's Wilbur Chocolate Company.) A very recent development is the formation of a partnership between Cargill, Bunge and CODEBA, a government agency, to invest in a grain terminal in the port of Ilheus.

Cargill's coffee operations, including those in Brazil, began with the acquisition of Dutch coffee- and cocoa-trading firm ACLI in 1983. Cargill also has significant coffee interests in Colombia, where Cargill Cafetera of Colombia has five mills in the heart of Colombia's coffee-growing region. It sold its worldwide coffee-trading business in 2000 to Ecom Agroindustrial Corp of Switzerland.

Indústria de Balas Florestal, one of the largest candy manufacturers in Brazil, used our pilot candy plant in Mairinque to test various formulations of Balinha do Coração, a new type of candy. The plant allows customers to test the production of candy, sweets and jams to lower cost and improve efficiency.[133]

Almost all of these activities are now strongly influenced by the magnitude of the growing soybean industry (and its lobby) in Brazil and Argentina and the engineering and construction projects intended to make it possible to get soybeans from the interior of the continent into the global market by means of *hidrovias*, or 'water highways'. The model, of course, is the Mississippi River.

Soybean Production

A map of South America in the October, 1995 edition of the National Geographic Atlas of the World indicates that only rubber and cattle constitute identifiable 'land use/land cover' in the interior of South America, with soybeans indicated only in the south of Brazil and a small area in Argentina north of Buenos Aires.

In the 1970s, anchovy fisheries off the coast of Peru collapsed, and this contributed to the use of soybeans as a substitute for fish meal in animal feeds in North America and Europe. In addition, a drought in North America led to a temporary suspension of shipments to Europe. The resulting increase in soybean prices led to rapid expansion of mechanised soybean cultivation in the southern Brazilian State of Paraná. A frost in southern Brazil in 1975 also speeded abandonment of coffee. Other factors inducing landholders in southern Brazil to switch from labour-intensive crops such as coffee included increased rights given to sharecroppers under a 1964 land statute and minimum wage laws that increased the cost of hiring labourers.[134]

In an editorial, *Milling & Baking News* pointed out that the crop area of Brazil and Argentina, at 419 million hectares, equals that of the US, but in addition, there are another 200 million hectares in the interior of Brazil that are being turned into agricultural land. This area has a more temperate climate than that of the corn/soy belt of the US with both longer growing seasons and the possibilities of double cropping. The Economic Research Service of the USDA, cited by the editorial, says: 'The potential for further growth of South American field crop output, if realized, could have profound implications for global trade and US farm exports, prices and incomes.'[135] What is most interesting in all such commentaries and reports is the absence of any mention of the major players in both North and South America, such as Cargill, that benefit whichever way the crop moves and at whatever price. The major hindrance to vastly increased agricultural production and export, however, is transportation. Apart from southern Brazil and the south of Argentina, the growing areas are a long, long way from an ocean, and trucking overland is very expensive. The solution? The transformation of the rivers into industrial waterways like the Mississippi, or *hidrovias*.

Thirty-five years ago the small farmers (sharecroppers, tenants, or squatters with less than 50 hectares of land) in the southern Brazilian states of Paraná, Rio Grande do Sul, and Santa Catarina grew coffee, beans, corn, and cassava. Then soybeans took over, rising from practically zero to 6.9 million hectares in 1980. After 1980 the area devoted to soybeans contracted while soybean production took off in the Cerrado of central Brazil. The term Cerrado refers to a characteristic set of vegetative types that include natural savannas and woodlands dominating 1.5–2 million km^2 in Brazil's centre-west states of Mato Grosso, Mato Grosso do Sul, Goias, and Tocantins and in parts of Bahia, Maranhao, Minas Gerais, and Piaui.[136]

In 1973, the Federal government created the Brazilian Agricultural Research Corporation (EMBRAPA) which began to develop soybean not only for the southern states but also for the vast tropical Cerrado. It was, ironically, assisted in this by The International Soybean Program at the University of Illinois, financed by the Agency for International Development. While US government financing of soybean development might have been good for some Brazilians, the greater beneficiaries, surely not by accident, are the soybean traders and processors – none other than Cargill, Unilever, ADM and Bunge.

Historically, the Cerrado had a low population density and large unoccupied areas, dominated by extensive cattle ranches, but the new soybean varieties, public road construction, and subsidized credit, fuel, and soybean prices changed that. The total annual crop area in the centre-west rose from 2.3 million hectares in 1970 to 7.4 million hectares in 1985, while the soybean area soared from only 14,000 hectares to 2.9 million hectares and then reached 3.8 million hectares in 1990. Heavily capitalized farms with between 200 and 10,000 hectares grew most of this.

Without the new soybean varieties developed by EMBRAPA, soil treatments (lime in particular), and machinery, the rapid spread of soybeans into the Cerrado would have been impossible, but credit subsidies were an essential precondition for the rapid adoption of agricultural machinery and soil amendments. Between 1975 and 1982, for example, one subsidized credit programme made $577 million in agricultural loans, 88 per cent of which went to farmers with over 200 hectares. Government subsidies combined with high international prices encouraged the spread of soybeans in Brazil, and this in turn increased the political power of the soybean lobby and enabled farmers and processors to obtain further government support.[137]

Where the Rivers Start

A 1986 World Bank document identified the eastern Bolivian lowlands as prime soybean land, and by 1995 there were close to 340,000 hectares under soya ... Around a third of the potentially rich 2 million ha. have so far been cleared for agriculture. Much of the credit for this goes to Joaquin Aguirre, who dreamed even in the 1930s of turning the Paraguay–Paraná River system into a South American Mississippi–Missouri. He finally achieved his dream of a port in 1989. Now Aguirre has signed two joint venture agreements, one with Cargill for expansion of the soya handling capacity, the other with Williams Energy of Oklahoma for grain, oil and diesel terminals on the Aguirre land. Meanwhile, efficiency on the *Hidrovia* is rising. With minor improvements and night navigation, the 45-day trip from Puerto Aguirre to Uruguay's Nueva Palmira transhipment facility could be halved.[138]

In 1996 Cargill and Central Aguirre Portuaria SA formed a joint venture to operate a grain, oilseed and oilseed-product storage and handling elevator in the Puerto Aguirre Free Zone in Quijarro, Bolivia. Bolivian soybeans were to be transferred from trains and trucks to barges for movement down the Paraguay-Paraná *hidrovia* system to Buenos Aires. The elevator would also serve Bolivian flour millers with wheat imported by barge from Argentina. Five years later it was reported that Cargill bought 51 per cent of the Puerto Aguirre grain port located on the Canal Tamengo in Bolivia close to the Brazilian city of Corumba on the Paraguay River.

The Paraguay–Paraná *hidrovia* was a project promoted by the Inter-American Development Bank and United Nations Development Programme (UNDP), which paid for and supervised its environmental impact and economic feasibility studies. The results of the studies were discredited almost as soon as they were made public because they pretended to be able to predict that only minor impacts would occur as a result of more than 250 heavy engineering works along 3,360 km of river in the world's largest tropical wetlands, the 350,000-km² Pantanal. As environmental groups and independent scientific experts pointed out, nobody (certainly not the consulting firms applying these hydrological models developed on channelized rivers in the US) really understands how the complex hydrological nature of the Pantanal functions.[139]

The latest plan for expansion of the Paraguay–Paraná *hidrovia* system involves the US company, American Commercial Barge Lines (ACBL), which plans to build a new port on the Paraguay River in the Pantanal wetlands 80 km downstream from Cáceres at Morrinhos in the state of Mato Grosso. American Commercial Barge Line is the largest barge operator in the US and its shipyard subsidiary, Jeffboat Inc., builds and maintains much of the country's barge fleet. (A 'barge' actually is a series of barges called a 'tow', lashed tightly together with steel cables and pushed by a 'towboat'.) 'We're pokey and low-priced', said the CEO of American Commercial Lines Holdings, the parent of ACBL and Jeffboat. According to its third-quarter 2001 report, American Commercial Lines LLC is an integrated marine-transportation and service company operating approximately 5,100 barges and 200 towboats on the inland waterways of North and South America. In South America, ACBL *Hidrovias* operates as a partner with Ultrapetrol SA in the joint venture UABL, SA. It is the largest barge line operating on the 3,600 km of the Paraguay–Paraná river system which covers Argentina, Bolivia, Brazil, Paraguay and Uruguay. UABL operates 331 covered hopper barges, 36 tank barges, 16 towboats and miscellaneous other equipment and facilities.[140]

Attempting to paint a 'green' face on its port project, UABL says that building the port downstream from Cáceres will eliminate the need for dredging and straightening the curves in the most crooked stretch of the river. Environmentalists counter that the official studies themselves recommended engineering works at more than 140 sites along the upper Paraguay to guarantee year-round passage of barges. In response to public pressure, and exposure of the total inadequacy and even illegality of many of the environmental impact assessments required under Brazilian law, the government has announced that most of the *hidrovia* projects have been cancelled. What this really means, however, is that those with an interest in these projects – such as UABL and Cargill – have simply changed their strategies and their arguments. They now push for specific projects which, taken on a one-by-one (case by case) basis, appear relatively harmless, but taken together they add up to the old *hidrovia* projects. This is certianly the case with UABL's port construction at Morrinhos.

Philip M. Fearnside, of the Department of Ecology, National Institute for Research in the Amazon (INPA) points out:

> Brazil's legal mechanisms for assessing environmental impacts and licensing infrastructure projects are incapable of detecting many of the most severe consequences of soybeans – especially the 'dragging effect' through which other destructive activities (such as ranching and logging) are accelerated by infrastructure built for soybeans. Even when problems are evident despite limitations of the environmental impact assessment system, the system is no match for the lobbying power of soy interests. In addition to the inadequacy of regulatory safeguards, the decision-making process that generates proposal after proposal for grandiose infrastructure projects is effectively disconnected from any consideration of the far-ranging impacts these projects cause.[141]

The expansion of soy monocultures in central Brazil has had its impacts in the central US, of course, and soy trading companies and shippers are calling for an expansion of the upper Mississippi locks and dams 'so that US companies can compete with the Brazilians'. ACBL's expansion plans, and the increasing domination of the South American soy business by US agribusiness giants Cargill, ADM and Bunge, demonstrate that the multinationals are adept at an old game – playing both sides against each other.

The Paraguay–Paraná river system is not the only one in South America, of course. There is also the Amazon, and the maze of rivers and wetlands that drain into it.

> Itacoatiera, a little-known Amazon River port (some 1200 or more km west from the mouth of the Amazon River), may expand to a major grain shipping point within the next few years, cutting the

transportation cost of Brazilian soybean exports to Venezuela or Europe ... Grupo Maggi and the government of Amazonas state have agreed to spend $29 million to make the river port accessible to grain ships up to 60,000-dead-weight-ton capacity by September 1996. Associated investments include a fleet of tugboats and barges, a 97 meter pier, a 70,000 metric ton storage silo, and a $21 million soya processing plant to produce soya oil and meal. These investments will, in turn, stimulate development of poultry, cattle feedlots, and fish farms. Most grain from the Mato Grosso state is now transported by truck more than a thousand miles to distant coastal ports, at a cost of $110 per metric ton and 11 days of time. The new route would cut transportation time to eight days and cost to $75. Nearby Venezuela is one of the world's biggest soybean importers and currently buys most of its soy products from the United States. The new port would increase Brazil's cost advantage.[142]

After the Itacoatiara terminal opened in 1997, 145 truckloads of soy a day started arriving at Porto Velho to be transferred to barges to go 800 km down the Madeira River to Itacoatiara where they are stored and loaded on ships for export. This new export route has cut the transport cost by a factor of three. Another soybean terminal at Santarém, Pará, began operation in May, 2000. Blairo Maggi, senator from Mato Grosso and head of the Maggi Group, has been financing soy planting in Santarém but it is not clear whether the new soy terminals at Santarém and Itaituba are being built by Maggi, Cargill, or both.[143]

Argentina

The Paraná River wanders all over, but from the south-western tip of Paraguay it flows directly south to Buenos Aires and the Atlantic Ocean through Argentina, where Cargill has been active since 1947 and is now the country's leading exporter of agricultural products. In 1995, Cargill Argentina reported annual sales of more than $1 billion, with about 80 per cent accounted for by foreign sales. The company says the presence of its FMD helped make the enterprise profitable by buying and selling currencies and other financial instruments while the commodity divisions bought and sold beans, grains and fertilizer. Cargill itself makes the point that its FMD has been able to learn from its branch in Argentina how to make up in the financial markets 'the money we would normally expect to make from the commercial side' during inflationary times. As a result, the Argentine FMD of Cargill 'developed a reputation for gutsy financial trading that resulted in winning big'.[144] (In contrast to Cargill's Brazilian website, information on its Argentine website is very limited and very dated: under 'news' the most recent item is 1997!)

Argentina *is* beef, and so is Cargill, but Cargill's notion of beef production is not what the Argentinians were doing. Cargill has said that it intended to transform the traditional practice of the cow–calf operators in the north-west of shipping their calves to the pampas for fattening on grass into one of fattening cattle in feedlots. More control can be exercised this way, more manufactured feed sold, and more dependency created. It also clears the land, so to speak, for soybeans.

In 1979 Cargill built a soybean crushing plant and a private port and terminal elevator at San Martin, near Rosario on the Paraná River, about 250 km north of Buenos Aires, to serve the soy growing region to the south. Cargill also built country elevators in the same region to funnel grain to its $24 million export elevator at Bahia Blanca, Argentina's best deep-water port to the south-west of Buenos Aires. Cargill has a malt plant there as well. It already had a soybean-crushing plant on the coast above Bahia Blanca at Necochea, which is also the site of a warehouse that receives fertilizer from Cargill's Florida phosphate plants. Farmers delivering soybeans to the crushing plant can return home with a load of fertilizer. Out of its 'concern for the environment', Cargill also provides support to a foundation that educates farmers about 'ways to conserve soil through modern farming techniques'.[145]

In 1996 the company expanded its Puerto San Martin soybean plant, making it not only the largest of its own plants, but one of the largest oilseed-processing facilities in the world. It also built a barge terminal facility so that it could load ocean vessels with soybeans and protein meals coming from north-west Argentina, Bolivia, Paraguay and Brazil, as well as providing an alternative source of raw materials for Cargill processing plants. At the time, Cargill clearly already had the Paraguay–Paraná *hidrovia* in mind.

Next came construction, in 1998, of a $14.4 million fertilizer port facility, the largest on the upper Paraná River adjacent to its large oilseed processing complex at Puerto General San Martin (also known as Quebracho). The facility includes a warehouse and high-speed bulk and bagged handling systems. Its location enables it to source the lowest-cost raw materials, whether it is DAP from Florida, Tunisia or the FSU, or urea from the Caribbean, the Middle East, the FSU, Brazil or potential domestic sources. Cargill also has a long-term agreement with Nidera SA under which each company agreed to handle fertilizers for the other. Cargill agreed to utilize a portion of its port terminal at Puerto General San Martin to handle fertilizer products for Nidera in the upper Paraná River region while Nidera agreed to use its port facility at Necochea to handle fertilizers for Cargill in that area.

Cargill diversified in Argentina as everywhere else and developed businesses that served its global operations. Juices and peanuts are good examples. In 1989 it established its Argentinian Juice Division to process apples and pears at Neuquen, in the valley of the river by that name in the west of the country. In 1997 it added a $6 million peanut-shelling plant in Alejandro Roca, Cordoba Province, to supply the domestic market and to provide it with a counter-seasonal supply to complement its Stevens Industries peanut business in the US. Cargill also built a malt plant in Argentina to serve the local brewing and agriculture industries. In 1995 it entered the Argentine flour-milling business with the acquisition of Minetti y Cia SA and in 1999 Cargill SA and Molinos Rio de la Plata SA merged their Argentine flour-milling operations to form the country's largest flour-milling company. As part of the deal, Molinos agrees to buy all its flour from the joint venture, which is 65 per cent owned by Cargill.

Venezuela

Cargill began operations in Venezuela in 1986 when it formed a joint venture and then purchased pasta-maker Agri-Industrial Mimesa. Three years later the company purchased a Pillsbury pasta and flour plant near Caracas, giving it two more brands of pasta, each targeting a different consumer group. Since Venzuelans eat more pasta than people in any other country except Italy, it was a good place for Cargill to gain experience with branded consumer products. It is now the biggest pasta company in the country, and given the similarities of Cargill's highly visible consumer-products businesses in other South American countries, it is obvious that the company has also been pursuing a regional strategy of visibility in the marketplace.

With a flour mill attached to each of its three pasta plants, Cargill is the second largest flour miller in Venezuela. It also sells cooking oils in Venezuela, having acquired the oil refining, marketing and distribution business of Ormaechea Hermanos CA in Turmero, Venezuela, in 1991.

Venezuela is now also the location of a new global-scale solar salt operation. This story is found in Chapter 18.

15 Juice

Next time you have a glass of orange juice, you might pause to consider Cargill's likely role in getting it from the tree to your table. That glass of orange juice, traced to its origins, can provide many insights into the customs and strategies of Cargill, both old and new.

Cargill's entry into the orange juice business arose out of its livestock feed business in Brazil, which was based on citrus pulp pellets purchased from Brazilian orange processors Cutrale and Citrosuco. In addition to using the citrus pellets in the feed it sold in Brazil, it was also shipping pellets to Europe as a component of the feeds it produced there. Then in 1975, when orange juice prices were lower, Cargill decided to make a move upstream, buying an idle frozen orange juice concentrate (referred to in the trade as FCOJ) plant in Bebedouro. The owner lacked operating capital. Once again, Cargill entered a new line of business by acquiring an existing facility for a low price. Its timing couldn't have been better. A disastrous frost in Florida pushed prices up and created a demand for Cargill's FCOJ.

At the time, about 80 per cent of Brazil's citrus industry was controlled by Cutrale SA and Fischer SA (later purchased by Dreyfus) and only about 5 per cent of its orange juice production was consumed domestically, the rest of it exported as frozen concentrate in 55-gallon drums (frozen after being put in the drums).

Cargill Agricola (the name it uses in Brazil) continued to follow its proven strategy of reinvesting earnings in additional capacity and increased output by 15 per cent annually. In 1980 it built a special juice terminal in Amsterdam to receive FCOJ in bulk and outfitted a container ship with a refrigeration system and a stainless steel hold to transport the frozen concentrated juice in bulk from Brazil to Amsterdam. A year later this same ship made its first delivery of 4.5 million litres of FCOJ from Brazil to the US.

By 1984 Cargill was shipping 90,000 tonnes of FCOJ out of Brazil annually – 15 per cent of the country's total orange juice concentrate exports. About a third of this was shipped to tank warehouses in the port of Tampa, Florida, from where it was delivered to customers such as Procter & Gamble, Kraft and Quaker. The rest of Cargill's exports went to Europe via Amsterdam. In 1985 Cargill Agricola built a second processing plant at Uchoa, and between them the two plants process about 45 million 90-pound boxes of oranges per year. This is 170,000 tonnes per year, at the rate of 40,000 oranges per minute, out of Brazil's total production.

About then Cargill developed a means of moving the concentrate at –7° to –8°C, meaning that the concentrate was still viscous enough to be pumped from tank to tank. Previously the concentrate was transported at –20°C. The new process made it possible for Cargill to concentrate the juice at its plants located at Bebedouro and Uchoa, pump it into a tanker truck, haul it to its terminal at Santos port (the port of São Paulo), pump it directly into its waiting bulk tanker and ship it to its Amsterdam terminal or to its terminal in Port Elizabeth, New Jersey, which it built in 1986. From these terminals it is finally pumped again into tanker trucks for delivery to dairies, food processors and institutions for bottling, blending, and food manufacturing.

Along the way, Cargill Agricola had Hyundai in Korea build the first vessel specifically designed to carry bulk orange-juice concentrate and peel oil. The 13,000-ton ship carries 11,000 tonnes of concentrates.

At the time, Citrosuco Paulista, Sucocitrico Cutrale and Cargill Agricola controlled 80 per cent of Brazil's orange-juice exports and the same three companies accounted for 53 per cent (800,000 tonnes) of the world's total orange-juice production. Brazil supplied about 40 per cent of the orange-juice market in the US and Coca-Cola (Minute Maid), Procter & Gamble (Citrus Hill) and Beatrice (Tropicana) were the three largest US distributors of this juice.

In 1986 Cargill worked out a deal with the São Paulo state railway in Brazil to minimize travel and turnaround time between Cargill's inland facilities and the port of Santos (the port of São Paulo) where Cargill built a private bulk terminal on land leased from the state port-operating authority. The result was, in effect, a private export corridor expected to move 440,000 tons of citrus meal, soybean meal and soybeans annually. The terminal facility was subsequently expanded to handle bulk and bagged sugar and in 2001 Cargill sold a 50 per cent stake in the terminal to local sugar group Crystalsev Comercio e Representacao Ltda, giving Crystalsev logistical control over the terminal which can handle over two million tons of sugar a year.[146]

In a break from its customary strategy of avoiding primary production, Cargill owns four citrus farms in Brazil totalling 10,000 hectares. The largest – Vale Verde – with 1.37 million trees on 5,300 hectares of reclaimed ranch land and tended by some 400 employees, is probably the largest single orange grove in the world. Cargill hoped that eventually its own groves would be able to supply 25 per cent of its needs in Brazil. Now, however, most of the oranges the company processes come from 'independent' growers with whom Cargill contracts to 'own' the trees for the season. At harvest time, the company hires 4,000 pickers and moves them from grove to grove. Because of the location of the orange groves, it is possible for Cargill's contractors to drive around and pick up day-labourers in the nearby cities, returning them after a day's work. This provides Cargill with the lowest-cost and most flexible labour force possible and virtually eliminates any kind of labour organizing.

A by-product of orange-juice processing is peel oil, which is used to add flavour and aroma to juice as well as in the production of cosmetics and pharmaceuticals. The final remains of the peel are then turned into pellets, 99 per cent of which goes to the European Union for animal feed.[147]

As a company, Cargill has always been willing to seek new opportunities and when necessary, is quite willing to enter into joint ventures with companies that are otherwise its competitors. In 1987, Citrosuco Paulista and Sucocitrico Cutrale, along with Cargill Agricola and Tetra Pak (the Swedish aseptic packaging giant), reached an agreement with the Russian government for a joint venture to process, package and distribute orange juice imported from Brazil. The agreement called for a plant to be built 500 km south of Moscow in Lipyceck and Soviet apple juice was to be bartered for Brazilian orange-juice concentrate. The plant was to process locally grown apples into juice for sale in Europe and the US and the hard currency earned then used to pay for the Brazilian orange-juice concentrate.[148] Cargill's website indicated, in 1997, that the company was processing FCOJ in Russia.

While Europe and the US remain the primary destinations for Brazilian orange juice, the growing affluence of the Japanese consumer naturally attracted large TNCs such as Cargill that saw new opportunities and became determined to force their way into the Japanese market. For more than a year before the April, 1992, opening of orange-juice imports, leading Japanese trading houses were looking for partners among overseas producers. At the same time, Citrosuco Paulista SA and Sucocitrico Cutrale SA of Brazil built storage for imported frozen concentrate in Toyohashi City, Aichi

Prefecture. The facility is able to store 20,000 tonnes of concentrated juice, a quarter of Japan's annual orange-juice consumption.

Being uncharacteristically late in entering the fray, Cargill was said to be interested in bypassing the Japanese trading houses and supplying juice directly to Japanese bottlers. The company got as far as choosing Funabashi, Chiba Prefecture, as the site for storage tanks, but that site ended up being used for a short-lived beef-processing project. Instead, in 1993 Cargill Japan began importing FCOJ from Brazil in 55-gallon drums for processing and blending with other juices, such as apple and pear, at its new 3,500-tonne capacity plant in Kashima, southern Japan.

In addition to looking for new markets, Cargill also began looking around for alternative sources of orange juice as soon as it knew its Brazilian operation was a success. It ended up in Pakistan where it built the country's first fruit juice concentrate plant in 1990 after four years of negotiating. Cargill had actually established its presence in Pakistan a decade earlier with the purchase of the British cotton trader Ralli Bros.

Cargill's Pakistan plant is located in Sargodha, Pakistan's main citrus area north-west of Lahore in Punjab. The Pakistan government provided incentives such as an eight-year tax holiday and duty-free import of all the machinery installed in the plant. The project was modelled on Cargill's plants in Brazil, and with the help of the US government, Cargill took a group of farmers from Sargodha to Brazil to learn how to improve growing techniques.

Besides broadening Cargill's supply base, Pakistan was attractive as the world's major producer of kinno, a satsuma-type orange developed at the University of California in the 1930s. The most important characteristics of the kinno are its taste, high vitamin C content and its colour. The frozen concentrate is exported to Amsterdam and used for blending and colouring because it has the rich orange colour of Fanta which until now has generally been produced artificially. Cargill believes the west and Japan are willing to pay the extra cost to have natural rather than synthetic colouring agents.

Cargill also acquired apple- and pear-processing facilities in Argentina and Chile in 1989 which together can produce 16 million litres of apple- and pear-juice concentrate per year, making Cargill South America's largest apple-juice exporter. Passion fruit and pineapple juices are processed into frozen concentrate at its plant in Fiera de Santana, Brazil. The company also has an apple-processing plant in Ferrara, Italy. Apple juice concentrate offers a big financial advantage over orange-juice concentrate in that it can be stored safely at chilled temperatures and even at room temperature for a period of time, though it would then have to be sent back through the line to be purified and/or reconcentrated before sale.

Cargill further extended its control of orange juice with the purchase, in 1992, of Procter & Gamble's large fruit-juice processing plant in Frostproof, Florida, which it immediately expanded. The plant's extensive cold storage capacity was particularly attractive to Cargill and in conjunction with its Brazilian plants, provides the company with the ability to supply orange-juice concentrate throughout North America year-round. This was certainly not a risky move for Cargill since for years it had been a supplier of Brazilian juice concentrate to Procter & Gamble, and as mentioned previously, Cargill-originated juice can be found in a wide variety of well-known brands as well as in private-label brands. In addition, much of Cargill's concentrate, like that of its Brazilian competitors, never reaches the consumer at the retail level as straight orange juice at all, going instead to bottlers and blenders to produce a great variety of soft drinks. In its first year of operating the Frostproof plant, Cargill processed about 7 per cent of the Florida orange harvest, including oranges from its own groves. (About 90 per cent of Florida's orange crop is processed for juice, while most California oranges are eaten directly.)

In conjunction with its phosphate-mining operations just down the road from Frostproof, which require holding large areas of land both as mining reserves and as reclaimed land, Cargill owns more than 1,200 hectares of orange groves in central Florida. There the company is doing some creative experimentation with orange production on reclaimed land.

Cargill appears to take the same thoughtful approach to waste management and ecology in the Frostproof operations as it does in its nearby phosphate-mining operations. It purifies and utilizes the processing water – including much of the 80 per cent water in the orange itself – by piping it 9.5 km to where it can be used to irrigate forage grass which absorbs the nutrients in the water. The grass is then harvested several times a year and fed to cattle, while the water that percolates through the ground is collected in drainage pipes and used to irrigate a 9-hectare citrus grove.[149]

Coming back full circle to Cargill's beginnings in the citrus business, the citrus peel from the Frostproof plant is itself ground, dried and pelletized, then shipped to Cargill's port facility in Tampa from where it is exported to Europe for sale as livestock feed through Cargill's distribution system.

Having integrated vertically just about as far as possible, in 2000 Cargill expanded horizontally, forming a joint venture with Florida-based SunPure, the largest grapefruit-juice processor in the world, to combine their North American citrus-processing businesses. Cargill is the general partner of the venture, responsible for day-to-day management. The alliance will create the largest non-branded citrus processor in North America, manufacturing both concentrate and not-from-concentrate as well as other citrus products such as oils and essences. Being non-branded means that they will appear under a host of store and private-label names.

Before forming the full joint venture, Cargill and SunPure had formed a smaller joint venture, Natural Cloud, to manufacture and market a functional beverage ingredient made from citrus peel. According to a Cargill Citro America specification sheet:

> Citro Pure Cloud is an all-natural clouding agent for carbonated and non-carbonated low juice and non-juice drinks. Functional characteristics include turbidity (the ability to give the liquid a more opaque look) and neutral flavour comparable to other cloud systems such as glycerol esters of wood rosin, bromated vegetable oil and sucrose acetate isobutyrate.

Cargill also formed a 40/60 joint venture with Empresas Iansa, Chile's largest agribusiness company, to produce concentrated juices. The new company, Patagonia Chile, produces apple, pear, raspberry and kiwi juice as well as apple essence, utilizing the former Cargill-owned apple juice plant in Rancagua and the former Iansa facility in Molina. It will be the largest non-citrus juice processor in the southern hemisphere.

With its range of both citrus and non-citrus juices and concentrates, Cargill is well positioned to be the reliable year-round supplier to the largest of retailers, such as Wal-Mart and Ahold of the Netherlands.

As both a producer and a global marketer/trader, Cargill enjoys a distinct advantage over companies operating as only one or the other. Cargill is well equipped to speculate in the commodity which it is also producing, and what it might lose as a speculator can be made up as a producer and vice versa – if it hedges its bets well. FCOJ is traded on the cotton exchange (it had to be put somewhere!) and the volatile FCOJ market is just what speculators love.

The system works in the same way as the futures trading described earlier. On the basis of its actual production, Cargill can sell a contract for delivery of orange juice at a future date at a specified price to a speculator. The speculator (sometimes called an 'investor') might hold the contract until exchange-rate fluctuations provide an opportunity to make a profit. Cargill, knowing that there has been unreported frost damage to the orange crop, may buy the contract back, knowing that the price of oranges for processing is to rise. At the same time, Cargill may be advising speculators to invest in orange juice futures, thus enabling Cargill to profitably sell the same contract again. Obviously this cannot go on for ever, but when the price goes down, Cargill juice can buy the contract back and sell the real orange juice instead of a contract for more expensive non-existent orange juice. It has been reported that every unit of a material commodity such as a tonne of FCOJ trades an average of 19 times before it reaches its final destination. This gives a company such as Cargill a lot of opportunities to extract profits from one shipment of FCOJ.

A rather engaging story appeared in mid-2000 about the origins of the production of oranges for juice in Brazil. The story is that it was US orange growers and processors that got Brazil into the business about 40 years ago – not long before Cargill bought its plant at Bebedouro – after one of the freezes that regularly devastates Florida's citrus crop (the Frostproof name notwithstanding). The Americans thought that the new groves they encouraged Brazilian farmers to plant in place of coffee would provide an alternate source of oranges in those years when frost wiped out their Florida crop. The groves in São Paulo state were so productive that within a few years Brazil had 45 per cent of the US market. By 1987, as a result of lobbying by the Florida growers, the US was imposing import duties as high as 63 per cent. The duties, together with a massive advertising campaign to get US orange-juice drinkers to equate oranges with Florida, managed to increase Florida's share of the US market to 64 per cent. The Brazilians did not take the market loss lying down, and countered by buying plants in Florida. Cutrale Citrus bought Minute Maid's plant, Citrosuco Paulista bought a plant and so did Dreyfus – and then there was Cargill. Now these four companies have the capacity to process 30 per cent of US orange production and the same four have more than 80 per cent of the Brazilian market.[150]

Fresh Fruits Spoiled

There are limits to how much orange juice anyone is going to drink and there are limits to how much of any single commodity one can sell in a single market. But if you have gained experience with oranges and orange juice, why not try other fresh fruit?

In August, 1991, after three years of scouting, Cargill purchased tree-fruit shipper Richland Sales Co. of Reedley, California, perhaps the largest peach shipper in the US, packing 5.5 million boxes of peaches, plums, nectarines, pears, kiwifruit and grapes from California and Chile. Less than a year later, Cargill decided it was safe to take a small expansionist step, and in May, 1992, Richland purchased Delano, California, grape-grower Prosper Dulcich & Sons.[151] Cargill officers described this in typical language: 'We wanted to start with something small to medium in size with the idea we could develop the business and get bigger.' Cargill almost immediately spent $1 million to install a 'superline' to double its daily packing capacity and accomodate different pack styles and brand labelling, thereby increasing its annual volume in 1992 by 10 per cent and its sales similarly to about $60 million. Doug Linder, as president of Cargill's Worldwide Juice, Fruit & Vegetable Division, was Richland's chairman and made all final decisions from Cargill headquarters in Minnetonka, reported *The Packer*. 'While Richland managers are given room to exercise their own judgement and style, all roads at Richland now lead to Cargill', said Linder.[152] Subsequently, Richland doubled the volume of grapes imported from Chile and got back into the apple and Asian pear business after a five-year hiatus. By 1996, Richland's fresh-fruit sales were falling fast and Cargill sold the entire business. According to *The Packer*, 'Some local shippers believe Cargill may not have been a good fit for the San Joachin Valley. "They brought the wrong attitude in ... This is a small valley made up of small growers and smaller parcels ... [Cargill had] a grain mentality or staple mentality or storage mentality. You can't have that mentality with a perishable crop like we have."'[153]

16   The 'Far East'

'We do it all the time, changing people's eating habits. It's good for them to have a bigger choice.' – Charles Alexander, director, US Agricultural Trade Office, Seoul, August 1994.

By the end of World War II in 1945, the US was the dominant power in the Pacific region and Japan, Korea and Taiwan became the victims of a new form of imperialism, US food aggression, as the countries were occupied by both North American troops and North American food. The food was sometimes in the form of direct relief to keep people quite literally from starving; sometimes it was aid for reconstruction, buying allies in the struggle against the 'communist menace'; and often, and more enduringly, it was creating a market for surplus American food. The net effect was the same: change the eating habits and create dependency on American supplies of wheat and white flour for direct human consumption primarily in the form of white bread, and create an intensive livestock industry based on imported feedstuffs, namely corn and soybeans from the US.

Once reconstruction and the industrialization process were under way, the US programmes were carried out under the 1954 Food For Peace banner of PL 480 which was essentially a market expansion programme wrapped in the American flag of anti-communism. Even as the threat of 'communism' was dissolving, as we saw in the discussion of the grain trade, US food imperialism was renewed and extended with the 1985 Farm Bill and the accompanying EEP. This programme was designed by US agribusiness interests to maintain US corporate domination of global food trade at any cost, particularly in the face of subsidized exports from the European Community/Union.

The most obvious victim of this food imperialism has been, and will continue to be, the domestic agriculture and rural society of the recipient countries. That has, however, never been an issue of concern to US-based TNCs like Cargill which have regarded Japan, Taiwan and Korea primarily as markets for food and feed. As one of the major suppliers of food and feedstuffs, Cargill has been both agent and beneficiary of this form of imperialism.

Taiwan

Thanks to very good contacts in Taiwan and South Korea, I was able to talk to a wide range of people in 1994 who provided a more detailed account of Cargill's activities in each country than is usually possible. Though somewhat dated this chapter nevertheless provides unusual insights into the ways of the Invisible Giant.

Taiwan long occupied a special place in the heart of American Cold Warriors as the last bastion of freedom confronting the Red Menace of China. As such the illusion of independence has been maintained even while the economic independence of the country has been undermined by the imperialism of development aid and the market demands of American agribusiness. In recent years the magnetism of the real China, with its billion potential consumers, has sidelined anti-communist ideology. As a regional hub and staging point for the great China market, however, during the 1990s Taiwan had a particular strategic significance.

The land area of Taiwan is 36,000 km^2, but the interior is mountainous and stunningly beautiful, as its name 'Formosa' attests. Only about 29 per cent of the total area, around the perimeter of the island, is suitable for agriculture. Four million of the island's population of about 21 million were defined as 'agricultural' in the mid-1990s, but there were fewer than 800,000 'farm families' with an average holding of 1.1 hectares. 'There are still too many people engaged in agriculture', stated the Council of Agriculture, or agricultural ministry.

Taiwan was the fastest growing market in the Pacific Rim for US agricultural exports according to *Milling & Baking News*, which reported: 'as in the case of South Korea, the modern wheat foods industry of Taiwan dates from shipments of food aid wheat and flour from the United States during the 1950s. Taiwan was a major beneficiary of the PL 480 programme until 1965. Since 1973, Taiwan has signed consecutive five-year wheat agreements with a group of participating US grain exporting companies.' The result was that the US accounted for about 88 per cent of Taiwan's wheat imports.[154] Taiwan also imports almost all of its soybean, wheat, sorghum and cotton requirements.

Until 1993, the Taiwan Flour Mills Association, representing all of Taiwan's 35 flour millers, was the sole wheat-purchasing agency, and it was encouraged by the government to purchase mainly from the

US, which meant that it dealt with US Wheat Associates as the representative of the US exporters, among them Cargill.

After 1952 meat output in Taiwan increased tremendously, with pigs replacing rice as the top-value agricultural product of the island and poultry in third place. The intensity of pig production has created a disease and manure problem and as a consequence, the Taiwan pork industry has gone in search of new territory to fill with pigs and pig manure, with Cargill's assistance, no doubt.

Cargill began trading in both Taiwan and Korea in 1956, the year Tradax was founded by Cargill as its global trading subsidiary. Cargill made its traditional physical entry in 1968 when it received permission from the Taiwan government to build a feed mill, which opened in 1971 in Kaohsiung. Cargill's second mill, in Taichung, opened in 1975. Each mill has a capacity of 30,000 tonnes per month.

While taking me through the Kaohsiung mill in 1994, superintendent Lin asked if we had noticed the amount of manual labour, a lot of it bagging feed and handling it by the bag. He said that headquarters in Minneapolis would like the plant to automate and get rid of all the manual labour, but the plant management has resisted and has tried to explain to Cargill that labour relations are too important to be treated that way. Lin said it is their own policy to automate slowly as people leave or retire. A lot of the employees have worked there since the plant opened 19 years ago, and the managers clearly felt some responsibility to these people.

Cargill might treat its regular workers well, including being concerned about their safety in the mill, but before entering I saw a peasant woman, barefoot, wearing only simple clothing, carrying a backpack sprayer to the mill building. When we came out, there she was, spraying the grass around the mill. There was certainly no regard manifested for her health and safety, but then she was probably just hired by the day to do such work.

Andy Chu, director of Cargill's Da-Tu mill in Taichung, did not quite share all the views of more central players in the Cargill hierarchy. Speaking as a Taiwanese, he was quick to tell me that the GATT was not good, that it would be bad for Taiwan farmers. With high land prices and high labour costs, Taiwan will become another Singapore, not a positive development from Chu's perspective. Chu also commented that with its excellent management skills and the volume of its purchased ingredients, Cargill's mills in Taiwan will continue to be successful, even as other companies leave for China.

While the mill management was resisting abrupt automation in the plant, it was also resisting head-office pressure to computerize all bookkeeping and bring it into line with the company's global system for all of its feedmills. The managers said they have very good staff who maintain close relations with their customers, and they do not want to jeopardize these good relations by automating them. It is important to talk with the customers, they said, and to know how they are doing. (As mentioned earlier, I heard exactly the same thing from the employees of a Cargill subsidiary in Nebraska, who did not like Cargill's centralized system either and said it would drive away their customers.)

Late in 1990 Cargill announced it would build a pork-processing plant in the Kojo region of Taiwan as a joint venture between Cargill Taiwan Ltd and Taiwan Sugar Corporation, which had been in large-scale intensive pig production on Taiwan since 1953. The plant was to slaughter 3,000 pigs a day (equivalent to the capacity of Japan's largest processing plant) when completed and export 30,000 tons of pork a year to Japan. Cargill figured that Japan imported about 360,000 tons of pork a year and if its plant operated at full capacity, it could supply just under 10 per cent of this quantity.

The plant opened in 1992, with its pig supply coming largely from its partner, Taiwan Sugar Corp, which was producing about 600,000 pigs a year in its own intensive pig-production facilities. The processing plant, originally a 50–50 partnership, is now owned 60 per cent by Cargill, 40 per cent by Taiwan Sugar, more in keeping with Cargill's standard policy regarding control. At the same time, it is clearly to Cargill's advantage to have a large minority shareholder with the kind of near-liquid resources and political power that Taiwan Sugar has. Gary Applegate, president of Cargill Taiwan Corporation, told me that 85 per cent of the plant's production was going to Japan as deboned chilled or frozen pork. (I had walked into his modest office on the third floor of a nondescript building first thing in the morning the day after a typhoon had knocked out most of the communications facilities, so our conversation was leisurely since, as he said, there was not the usual stack of faxes from the day and night before to deal with.) Applegate told me that the complementarity between Cargill's feed production and Taiwan Sugar's pork production was less than Cargill expected it would be, but if Cargill is the supplier of the main feed ingredients, imported corn and soybeans, it is a good bet that Cargill makes a fair bit of money even without doing the feed manufacturing itself. By the same token, if the margin on pig slaughtering and processing is a bit slim at times, Cargill has still made money on the feed components they have sold to Taiwan Sugar. As for their partner, Applegate credited Taiwan Sugar with being the best-run government company in the world – a very high level of praise, given Cargill's competence and experience.

Applegate also told me that in Taiwan, as in Korea, feed was their way into the country, implying that it would also be used in pursuing their traditional strategy of establishing a beachhead in a new location, such as China, where the company already had a joint-venture feed mill, which Applegate described as a learning experience. Cargill was already building two more mills in China and intended to have five feed mills in operation in China by 1997, with most of the feedstuffs expected to be of Chinese origin. (Unfortunately, I could not afford the time or cost to make the research trip that would have been required to update this.)

Applegate expressed his pleasure with his staff that are so good at buying soybean meal and other commodities that they can buy in the US mid-west for delivery in Taiwan and match the cost of meal purchased in Taiwan from much nearer sources in China. 'They know where all the soybean meal is, how much there is and where it is going, all around the world.'

Applegate was also pleased that a branch of Cargill Investor Services (CIS) had been established in Taipei in May, 1994. 'It lowers the cost of doing business by providing low-cost money [low-interest financing].' This is an interesting revelation because Cargill advertised CIS as handling only its customers' accounts, not its own. Cargill's directory also listed 'M.A. Cargill Trading Co.' in Taipei, and Applegate described this as a molasses-trading company that also handled other feed components, such as tallow, as well as hides and other low-value raw materials. M.A. Cargill also operated Sea Continental, a shipping company that owns no ships, but arranged cargoes for Cargill and others throughout the region.

In looking to the future, Applegate said Cargill Taiwan had reached a limit with feed and pork processing unless Taiwan became a regional hub for 'further processing' and distribution of meat produced elsewhere, such as beef from Australia, the US and Canada, and chicken from Cargill's plant in Thailand. In this scenario, the white meat would be shipped to the US where it is preferred, and the dark meat to markets in south-east Asia where it is preferred. Taiwan itself is too small to be an attractive market, compared, for example, to Korea with twice the population of Taiwan (three times if North

Korea is included). All that Taiwan has to offer is its geographical location and its skilled but high-priced labour.

Two days later, to my surprise, I heard this very same outlook from three policy researchers at the Institute for National Policy Research (INPR), which is funded by a wealthy Taiwanese financier specifically to create policy for the president of Taiwan. When I met with the researchers, I asked them what role the corporate sector in general, and companies like Cargill in particular, had in policy formation. Their very polite response was that they had never thought about it. Not only had they never given a thought to the role of agribusiness in shaping agricultural policy, they knew absolutely nothing about Cargill and what it was doing, but assured me that they could work in partnership with companies like Cargill to secure their own economy. I don't think they really expected me to believe that.

It was not all smooth sailing for Cargill in Taiwan, however. At one point, the company complained that it was being shut out of soybean tenders by the BSPA ('The Breakfast Club'), Taiwan's leading purchasing group, which was buying about two panamax cargoes a month of US soybeans by open tender. ('Panamax' refers to the maximum size ship that can navigate the Panama Canal.) Cargill's general manager in Taiwan at the time, Jason Lin, described the Taiwanese market as 'an oligopoly with only 15 major crushers'. Cargill usually got one-third to one-quarter of the business, but that changed when a key BSPA member company reported that a 54,000-tonne cargo sold by Cargill was significantly short-weight when it was delivered and the club demanded that Cargill and the carrier, Maersk Line, take the cargo back or offer compensation. Since they refused, the group continued to boycott Cargill until the dispute was settled.[155]

Referring to his experience of the US in trade negotiations, and reflecting the attitude of representatives of smaller countries, a senior Taiwan GATT negotiator said to me: 'So many companies, so many senators ... We do not distinguish between the US government and the companies. We can make no difference. We don't have the resources to counter the lobbying of the US Wheat Associates.' He continued: 'The rice millers force the US government to subsidize rice and force an extension of the market for it. So the US will impose policy on other countries like Taiwan. The US imposes its policy on Japan first, then Taiwan. If Japan yields to US pressure, we must follow.'

Curious about Cargill's partner, Taiwan Sugar Corporation, I interviewed three of their senior executives. The company was formed at
the end of World War II out of four Japanese-managed companies
that were originally established by the Dutch East India Company
in the seventeenth century. When Japan occupied Taiwan in 1895,
they recognized the importance of the companies and kept them
going, using the sugar they produced to make alcohol for fuel during
the war. In the late 1940s the Chinese came from the mainland and
took over the company, incorporating it into the Kuomintang
(KMT), the governing party of ruler Chaing Kai Shek. This resource
made the KMT the richest political party in the world, according to
one Taiwanese opposition member.

Since 1949, Taiwan has been a major sugar-exporting country,
though its exports peaked in 1953. Over the years Taiwan Sugar
diversified, but still owned 58,000 hectares of land in Taiwan and
sugar remained its primary business activity, even if it lost money.
The annual deficit could be made up by selling a few hectares of
land. Having accepted a limited future for itself on the island, Taiwan
Sugar began moving abroad, building a 6,000-tonne-per-day sugar
mill in Vietnam and exploring in Indonesia. It also moved its bagasse
pulp mill to Australia while importing cows from Australia to
develop its domestic cattle operations as a kind of holding operation
on its unprofitable sugar lands.

Then I had a call one day in 2001 from a farmer in Flagstaff,
Alberta, about 150 km south-east of Edmonton, wanting to know
what I could tell him about Cargill's relation to Taiwan Sugar which
was trying to get permission to build a very large pig-production unit
in the area. Taiwan Sugar had already been turned down in southern
Alberta and was making its second try to get permission. Given
Cargill's history with Taiwan Sugar and its feed milling operations in
Alberta, I think he was right in suspecting that Cargill was functioning as an invisible agent for Taiwan Sugar.

Korea

The stage was set for the present economic structure of Korea by
a series of unfortunate historical experiences: the 35-year Japanese
occupation; the forcible division of the country by the US and
USSR in 1945, leading to the extremely destructive Korean War
(1950–53); the corrupt rule of Rhee Syngman, which neglected
land reform and opened the south to the beginnings of economic

domination by the US; and finally the 1961 military coup by Park Chung-Hee who took it upon himself to modernize the nation. It is the Park model – rapid industrialization based on exports – which set the pattern of economic development for South Korea. The Park plan demanded a massive, low-wage factory labour force, which was obtained by a cynical neglect of the agricultural sector and a low-grain-price policy coupled with imports of foreign agricultural goods, resulting in the exodus of huge numbers of rural people who could no longer survive by farming. In 1990, 5 per cent of the Korean population held 65 per cent of the total private land, or 47 per cent of all land, and the majority of those owners were said to be chaebol [conglomerates], the top five of which are Samsung, Hyundai, Daewoo, Lucky-Goldstar and Hanjin. These big export corporations also benefited from massive loans at favourable rates, obtained with government backing from the US and Japan.[156]

It was not enough, apparently, that Korea was already the third largest importer of US agricultural products. The frequently seen ad for Marlboro cigarettes – the American Cowboy image – seemed to represent not only the attitude toward Korea of the tobacco pushers, but of Cargill, Continental, ADM, and the US government itself. It was particularly noticeable in the attitude of the Korea director of the US Meat Export Federation and the director of the Agricultural Trade Office (ATO) of the US Embassy both of whom I met with in Seoul.

The offices of the ATO of the USDA in Seoul were in a nondescript building behind the fortified US Embassy. Behind the double armoured doors, metal detection devices and armed guards of the ATO (long before September 11, 2001) were also the offices of the US Meat Export Federation, US Wheat Associates, and other publicly supported lobbies of US agribusiness. Besides standing on the American flag, these groups receive half their funding directly from the US government and the other half from the industries they represent.

The US is not well liked in Korea, though Japan is disliked even more for its decades of domination of Korea which ended only with Japan's defeat in 1945. The US is disliked for its long-time support of repressive governments, its aggressive attitude toward Korea as a market, and its hostility toward reunification of Korea. The US has, in fact, gone so far as to act as an *agent provocateur* creating continuing conflicts between the North and South, for example, the alarmism generated over the so-called nuclear threat from the North in the summer of 1994, which ended only when President Clinton did an about-face, and most recently President Bush's designation of North Korea as a member of his 'Axis of Evil'. In this case as well, before Bush actually arrived in Korea someone had told him to change his tune fast in light of the public anger he had aroused in South Korea.

Even with the correct address and a knowledgeable guide it took quite a while to find the Cargill office in Seoul, and one of the first questions Yoon Ik-Sang, director of Cargill Trading Ltd, Korea Branch, asked was: 'How did you manage to find this office?' I smiled and gave him an evasive response, and he smiled, acknowledging, with Asian courtesy, that I did not need to and was not going to inform him. I was later told by Korean farmers about Cargill's ads in the farm papers: a large blank space with no words, only the Cargill teardrop-in-a-circle logo in one corner, without the customary 'Cargill' inside. The company kept its name out of sight, while still trying to familiarize the farmers with its logo. This won't work with the company's brand new logo.

South Korea – 70 per cent covered by mountains, inhabited by 43.7 million people, and now highly industrialized as a result of deliberate state policy since 1960 – was increasingly attractive, not as a supplier but as a market for imported food and agricultural products. From what I was told, Cargill shared this perception of Korea's opportunities.

In 1994, South Korea's 2 million farm families comprised 10–12 per cent of the country's population, the average farm size was 1.2 hectares, and even that holding was often broken up into several small fields. My strongest image of the Korean 'rural' landscape is of these small, intense rice paddies abutting huge new apartment complexes housing probably 10,000–20,000 people each.

Cargill proclaims the benefits of agricultural 'modernization', but the reality is that Cargill, in Korea and elsewhere, has to engage in a balancing act between serving the interests of huge conglomerates like the Korean *chaebols*, the demands and sensitivities of a still numerous rural, if not farming, population, and its need to shape the business climate to suit its own corporate interests. These three agendas seldom coincide.

Nevertheless, the fact is that Cargill, as a commodity buyer, trader and distributor, could serve the interests of both the *chaebols* and itself, while as a processor it could maximize returns by processing commodities, such as livestock feed, and compete with the *chaebols*, as long as it did not threaten them. In the case of beef and pork, Cargill could play a more direct and uncomplicated role in the marketplace since it had sources the *chaebols* could neither reach nor control. Cargill's overall advantage was in its integrated functions, interests and flexibility. These factors enable it to get what it can where it can without being so aggressive as to antagonize its opposition into hostile action.

Cargill Trading was established in Korea in 1986, at a time when the company was expanding rapidly into new regions. Prior to that, Cargill acted only through local agents. At Cargill's Taipei office, 'Excel' was on the door, but I was told that it is a separate company with no connection, and when I asked where I could get information about it, I was referred to Brad Park of the US Meat Export Federation.

Cargill, according to Yoon Ik-Sang, was involved in the cotton trade in Korea through its subsidiaries Hohenberg and Ralli Bros. It was not involved in rice trading because in global trade it is too small and because the US Rice Millers Association has the market sewed up with the Korean government. (While Cargill became a rice miller in 1992 with the purchase of the largest rice mill in the state of Mississippi, it apparently had not yet made its presence felt in the Rice Millers' Association.) At the time, Cargill had only a small role in the juice market because Cutrale and Dreyfus, the major exporters of Brazilian orange juice to Korea, got there first. Cargill did not bother with the seed business in Korea, according to Yoon, because the market was too small. Livestock-feed-grain importing, feed milling, and oilseeds were the most public aspects of Cargill's involvement in Korea.

Cargill first got permission to enter the livestock feed business in Korea in 1986 when a process of modernization and consolidation was taking place in the industry. The domestic players agreed that Cargill should not be allowed to build a new mill and made their views known to the government. Cargill got around them by agreeing to buy an old mill, Young Hung Mulsan Co., in southern Korea. It then proceeded to dismantle the mill and move it to Chungnam in west-central Korea where it was rebuilt.

In 1970 Korea manufactured only .51 million tonnes of compound feed ('mixed with a shovel'), while in 1991, 11.5 million tonnes of compound feed was produced, utilizing 8.5 million tonnes of imported feedstuffs. This dramatic increase reflects both the increase in livestock numbers, and an increase in the use of manufactured feeds requiring imported feedstuffs. And of course Cargill was a major supplier.

While the Korea Feed Association described itself as a major force in public policy, it also said that Cargill was the major influence on US trade policy in Korea. Whichever way Korean livestock and meat policy goes, however, Cargill cannot lose. If domestic beef production declines, and with it the market for livestock feed, Cargill may lose as a feed importer and miller, but it will gain as a beef importer. Or vice versa.

The Korea Wheat Flour Milling Industry Association represents the three big 'tycoons' of Korean industry (Hyundai, Daewoo and Samsung) who were reborn after the liberation from Japan at the end of World War II. They were able to import very cheaply the raw materials for the 'three whites' – sugar, wheat flour and cotton – under the US Food For Peace and aid programmes, and process and sell them dearly in Korea. The 'tycoons' had supply contracts with companies like Cargill, according to Yoon, and the outcome was that the suppliers and the processors did very well at the expense of the Korean and American people. Samsung, in particular, made windfall profits as an importer and processor of US PL 480 wheat with Cargill as its agent. The US government provided concessional wheat sales to Korea under PL 480 until 1981.

In the mid-1990s the big Japanese trading companies controlled 90 per cent of Korean food grain imports (about 2 million tonnes of wheat per year) both for the *chaebols* and for the many small flour mills. Of course, it helps that the GSM 102/103 programme of the US government provides financing, which neither Canada nor Australia provide. And while the Japanese trading companies may control the trade, they still have to buy the grain, and it is companies like Cargill which have the originating capability (country elevators, etc.).

The same pattern could be found in soybeans, a major Korean food source almost entirely imported. The major soybean processors were all associated with one or another of the *chaebols* and bought from the major importers.

Cargill first sought permission in 1988 to build a soybean-processing plant to produce edible oil in Korea, a perfectly logical development for its worldwide integrated oilseed sourcing and supply system. A year later the Korean Finance Ministry gave Cargill permission to process 300,000 tons of soybeans annually after both Commerce Secretary Mosbacher and Trade Commissioner Carla Hills had pressed the Korean government to accept Cargill's proposal.

Korea's Agriculture Minister, however, said he would not approve the investment by Cargill because he feared that it would devastate the nation's soybean-related industries and soybean farming. 'If Cargill expands here, it won't be just a matter of the sale of a few more bags of fodder or tanks of oil. That is because we can see in this move the terrible plan to export raw materials as well as to process them here for profit, then ultimately to force the small Korean businesses into bankruptcy and take over the whole market.'[157]

Behind the scenes there was a slightly different account of the reason for Cargill's failure to gain permission. In February, 1988, Rho Tae Woo became president of Korea. His son-in-law was president of Dongbang Yuryang and according to the Korea Feed Association and others, Rho could not allow Cargill to threaten a family interest. With 90 per cent of Korea's essential soybean imports coming from the US at the time, the big three companies announced that they would not continue to buy from the US if Cargill was allowed in. While this story may not be true, it is certainly a true picture of the kind of power relations that Cargill both enjoys and has to deal with. It does not always get its way – at least not immediately.

At the end of 1992, after continuing pressure from the US government, the Korean government finally gave in to Cargill in spite of the strong opposition of Korean soybean growers, processors and farmers' organizations. According to the Korean Soybean Processors Association, once Cargill got approval, it discovered that there was already 25 per cent overcapacity in the processing industry, making any additional capacity an unwise investment. Recognizing Korea as a 'mature' market, with little room for expansion, Cargill started looking elsewhere.

As the export promotion arm of US beef packers, the US Meat Export Federation is responsible for boosting US beef exports, anywhere and everywhere, including forcing open various markets to US beef. It does this by pressuring members of the US Congress which pressures the US government which then pressures the Korean or other governments. Given that Korean cuisine tends to emphasize beef and that the country is the world's fifth largest beef importer and the third largest importer of American beef, it is not surprising that the heavy guns of the US meat industry have long been directed at the Korean market.

The *Korea Times* commented in 1988 that US Trade Representative Clayton Yeutter: 'demanded that Korea import US beef not only for tourist hotels but also for restaurants after the general elections

... [He] also demanded that Korea simultaneously lift its import ban on frozen potatoes used by McDonald's fast-food chain.'[158]

Brad Park, a Korean American, was the Federation's representative in Korea. He described for me the background to the present Korean debate about food imports, reflecting in his description the policy position of Cargill and others:

Korea was considered to be an agricultural country until about 1960. From the beginning of the Korean War in 1950, when almost everything got destroyed, until around 1960 everyone was kind of struggling to get something to eat and about 70 per cent to 75 per cent of the population was considered to be farmers, or of farm background. Then during the 1960s Korea started to industrialize and by 1990 the population in the farming sector had declined to 11 or 12 per cent. A lot of farmers' children had moved to the city, but remained farmers at heart. At election time, all these people become farmers again.

Now farm production comes to about 5 per cent of the GNP [gross national product], so economically it does not make sense for government to put money into agriculture. But politically, it does, because 50 per cent of the population will vote as farmers in election time. About 2000, or 2010, this will change, when the good old memories of farming are gone. There are a lot more important issues than agriculture.[159]

Beef was first imported into Korea in 1981 and thereafter imports both of beef and breeding cattle increased so rapidly that in late 1984 and early 1985 the government halted all imports of beef in order to curb the decline in prices. In 1988, using as an excuse the need to feed the American tourists flocking to the Olympic Games, the US brought a lot of pressure to bear on the Korean government to reopen the market. If they had to do it, however, the Korean government wanted to do it rationally, so they established the Livestock Products Marketing Organization (LPMO) to import beef for 'general' restaurants and manage the total until full liberalization of the beef market in 2001. It also set up the Simultaneous-Buy-Sell (SBS) system in 1993 to handle an increasing share of imports on the basis of negotiations between buyers and sellers.

Park described the role of the Meat Export Federation as that of advisor to the US government, but, he said, it is entirely up to the US government what position they take:

We are strictly a voice from industry. We – Wheat Associates, Feed Grains Council, Meat Export Federation, Soybean Association and so on – are called cooperators with the Agricultural Trade Office (ATO). We do not have to be here, but the ATO encourages us to because we are operating as a group. It is more convenient and economical for us to be here in this situation [sharing the same offices]. Funding is one half membership, one half government. We have about 300 members, and about 100 of them are packers. A lot of times we are invited to trade negotiations as observers. Our main function is as a liaison office, for the convenience of our members who want to do business here, and for Koreans who want to buy.

Park said that there is a very symbolic saying in Korean, written in Chinese characters, 'explaining' imported food products: 'Body and Land are not two different things', in other words, what you eat is what you are. It is a good idea, he said.

One thing we would have to say is that there are no Korean cattle in this country because virtually all the feed grain is imported. The way I like to interpret this is that when you are eating good quality food, you keep your good health and build a good situation, regardless of where it comes from. I think we have to see it that way. Now we have much more choice, we can pay a lower price and get good quality products because of this more open market system. The government should try to get the most high value, most economical food for the people.

As director of the ATO, in the corner office down the hall from Park, Charles Alexander was responsible for 'market development'. Dennis Voboril, who refered to Alexander as his 'boss', was US Agricultural Attaché.

Alexander explained that his job is to help Korean importers of US products and US exporters. 'We try to solve the problems that industry encounters – such as 37 containers – $1.3 million worth – of hot dogs held up at the container port because, they said – they discovered a regulation that they had not been enforcing – these are frozen cooked sausages and there is no category in the food code for this product, therefore they have to have a 30-day shelf life. With a 30-day shelf-life they are perfectly importable – for up to 30 days from date of manufacture. But it takes about 30 days to get them here.'

Two months after I heard that story from Alexander, the US National Pork Producers Association, the National Cattlemen's Association and the American Meat Institute filed a petition calling on the US government to impose trade sanctions on Korea in retaliation for halting trade in frozen cooked US sausages. The associations complained that Korea imposed short-shelf-life requirements and long inspection procedures to prevent American meat from entering the country.

The working relationship between the publicly funded US Meat Export Federation, the US Agricultural Attaché, and the commodity associations is very cosy. 'Here at the working level we don't very often stand back and look at policy', said Alexander. 'We're dealing on a day-to-day basis – we want access, we need access, our people want access, we should have it – Korea is free to export all its products to the US, we ought to be able to export our products here. That's what it boils down to for us.'

> In Korea, Japan, and Europe, there is a policy which says we want *x* number of people on the land. In the US the push for the past 50 years has been for efficiency in crop production. And you have seen the farm population drop to 1.9 per cent of the population, with 150,000 farmers producing 50 per cent of the production in the US. You can't get much more efficient than that. So when we approach the Korean market, our natural mindset is that they need to get bigger, there are too many Korean farmers. They have to be more efficient. It doesn't matter how good a farmer you are, you cannot make a living on 1.2 ha. of land. You can't sell enough product to make a living. You're not going to get any sympathy for Korea from me! Government functionaries listen to the people who pay their salaries – just about everywhere in the world except Korea and Japan. I've never met a more officious bunch than these damned Koreans I meet seem to be. They are bright, they are smart, but they are convinced that they are right.[160]

Expressing the traditional Cargill approach to other people's problems, after the 40 per cent devaluation of the Korean currency in 1997 amidst an economic crisis gripping the entire region, Dan Huber, president of Cargill's Asia Pacific Sector, said: 'In the midst of the crisis lie opportunities for Cargill. We're looking at increasing our investments in grain and food processing businesses.' Despite the economic turmoil, Cargill's Korean operations will exceed budget. 'I'm very proud of the way we worked together to navigate through the financial crisis in Korea these past six months', Huber said. Since the beginning of South Korea's financial crisis, Cargill has carefully tracked the location of every Cargill ship headed for Korea and the status of every letter of credit. Without a letter of credit, Cargill doesn't hand over the cargo. Before the crisis, Cargill held a 40 per cent market share of the Korean feed grain market. In addition to over 2 million metric tons of grain, Cargill also sells salt, hides, sugar, coffee and vegetable oil to South Korea. 'It's a very difficult time, but we use the difficult time to make money', said Woo Young Lee, president of Cargill's Korean feed business.[161]

Japan

> When William Wallace Cargill bought a small grain elevator in Iowa in 1865, Commodore Matthew Perry had already succeeded in inducing Japan to open its ports to US merchant vessels ... When Commodore Perry brought his black ships, the menacing symbol of advanced Western technology, to Tokyo Bay, the Japanese had no habit of eating beef. They did not need feed grains.[162]

The trade journal *Milling & Baking News*, in 1994, provided a brief but crucial insight into modern Japan's food history in an editorial:

> Prior to and during World War II, Japan was primarily a nation of rice-eaters ... Decisions made during the US occupation following the war led Japan to begin large-scale imports of wheat and to encourage demand for bread and other grain-based foods. A school lunch programme was started, providing children with bread or rolls every day. The success of this effort is one of the great tales of modern-day milling and baking ... Another legacy of World War II, though, is a domestic support program for rice that was originally put in place by General Douglas MacArthur to spur people to remain in farming in the hope of building a base of pro-democracy farmers.[163]

While the forceful conversion of the Japanese to wheat-eating may be viewed as a smashing success by the US industry that has benefited from it, Japanese themselves have other opinions. My host in Tokyo told me of her memories, as a schoolgirl, of being quite literally forced to eat the horrible white bread that was provided under the school lunch programme. They were told to hold their noses when they were forced to drink the 'milk' from skimmed-milk powder that had been so generously provided by the US. Then they could have their rice.

Japan was officially occupied by the US, with the administration of its government under the command of General Douglas MacArthur as Supreme Commander of the Allied Powers, until 1952. Although it had ordered the great Japanese business combines, the *zaibatsu* (counterparts to the Korean *chaebols*) to be broken up after the war, as the Cold War took over the US realized that it would be prudent to overlook the re-emergence of the *zaibatsu* in the interests of strengthening Japan's economic recovery. Such apparent concern for Japan's economy was a thin veil for the more powerful passion of anti-communism. The US preferred an anti-communist ally to a reformed society, and the *zaibatsu* to a socialist economy.

For example, although Mitsui was supposedly dissolved in 1945, by 1960 it had emerged as the world's largest trading company. By 1994, the top five *zaibatsu* dominated the corporate world.[164]

Cargill's life in Japan is an interesting story of Great Power relationships, and if Cargill has found its way hindered, if not blocked outright, by the *zaibatsu* of Japan, it is in striking contrast to its experience in India, where it was not the established powers but the powerless peasants who threatened its progress. These different forms of resistance to a corporate intruder stand at opposite ends of the spectrum of power.

As Cargill experienced, the five giant Japanese trading companies exercised significant control over the economy of Japan, if not the government itself. Cargill had to find its allowed, or tolerated, place in their social order, while hoping that the Japanese market would be forced open by a combination of corporate, US government and WTO pressure.

Cargill has been doing business *with* Japan since 1950 when the Foreign Trade and Exchange Law was put into effect in Japan and Cargill was able to act as an agent or buyer for Japanese interests, supplying grain from Kerr Gifford in Oregon. In 1956 Cargill started actually doing business *in* Japan when its subsidiary Tradax (then based in Montreal) acquired Andrew Weir (Far East) Ltd, an importer and supplier of foodstuffs that had been set up just after the war.

An informal history of Cargill in Japan, written by a retired employee, describes how, in the late '50s, feedgrains were 'low key' but foodgrain imports expanded rapidly and contributed to a transformation of Japanese lifestyle from traditional diet to western habits [white bread]. The same writer describes how a 'friendly crusher' built a new oilseed-crushing plant that could accommodate panamax vessels. 'Friendly crusher' turned out to be Fuji Oil Co. Ltd, an affiliate of Itochu (successor to C. Itoh), and their supplier was Cargill, the only company at the time that could afford to trade in whole shiploads of oilseeds. Cargill was able to guarantee supply in return for having a guaranteed buyer. At times this meant that Cargill was even able to buy from competitors and sell to a 'friendly crusher' at a handsome profit. (Itochu also has 'friendly relations' with another oilseed processor, Ajinamoto, and Ajinamoto and Cargill are business associates in Iowa.)[165]

Oilseed meal was another product Cargill was able to purchase and deliver by the shipload and consequently at low prices in the late 1960s and early 1970s, giving it control of more than 90 per cent of Japan's imports of oilseed meal, a key livestock feed ingredient. In 1973 Tradax made an after-tax profit of $100 million on its soybean trading alone, with Tradax Japan contributing more than 10 per cent of that. Later, when Tradax Japan nearly lost its shirt in the process of some fancy dealing to supply Itochu, its 'friendly' Japanese Trading Company (JTC) came to its rescue. The company did business as Tradax Japan until 1985 when the name was changed to Cargill North Asia Ltd (CNAL). In 1992 the name was changed again, this time to its current Cargill Japan Ltd.

The company's business in Japan expanded in 1972 when Cargill Inc. acquired C. Tenant Sons & Co. of the US, which had already established a Tokyo branch in 1963 through its subsidiary Tenant Far East Corp of Hong Kong. This branch carried on an export–import business in non-ferrous metals, which Cargill has continued to do while adding other lines of activity. 'At first, the company merely carried its products to Japanese ports, where they were turned over to Japanese companies for domestic distribution and sale. Cargill was doing business "to" Japan, in other words, not "in" Japan. The company would be much better off today if it had built its own facilities back in the late 1960s, but at the time ... our prime interest was in trading grains', said J. Norwall Coquillard, president of Cargill Japan.[166]

As the value of the Japanese yen began to climb in the mid-1980s, Cargill changed its approach and began to establish real beachheads. Its first effort followed its traditional practice of acquiring established businesses in which it already had expertise: it bought the grain business of Honda Motors in 1983.

In 1985 it followed its alternative strategy of establishing a new business building on one of its 'core competencies': Cargill's seed division established an experimental farm in South Kyushu. A year later a second experimental farm was established by Cargill Seeds in Tokachi region of Hokkaido and Cargill Japan's seed department, located in Sapporo, distributed hybrid corn and sorghum seed, 95 per cent of which was imported from Cargill subsidiaries ('principals') in the US and France since it had no Japanese seed lines.

There is no record of what happened to the grain business Cargill bought from Honda, but it was probably absorbed into its larger grain operations without giving Cargill a real beachhead in Japan. As for its second beachhead, the seed business, it remained a small operation until Cargill sold its entire international seeds business to Monsanto in 1998.

Japan was an affluent and growing market, however, and Cargill was obviously not content with simply being a trader and supplier to others or with a marginal role in one of its chosen lines of business. It still wanted to establish a beachhead in a major industry, such as livestock feed, where it could run its own business, establish its own customer base and grow. With Japanese imports of feed ingredients running at 16 million tonnes a year, it was a natural for Cargill.

Self-sufficiency, however, has been basic policy for Japanese agriculture since World War II, and after 1953 it was Japanese policy to encourage the expansion of livestock production while favouring Japanese companies. Thus, while anyone could build a feed plant as long as they met the building and milling standards, they had to obtain authorization from the Customs Office to import the feed ingredients duty free. This authorization was a ministerial function based on evaluation of supply–demand conditions in the area where the mill was to be built. Without being able to import feed ingredients duty free, no plant could compete in the Japanese market. At the same time, however, Japan was becoming increasingly dependent on imported feedstuffs.

When Cargill decided to break out of its confinement in 1985 it announced that it would focus on beef production and packing on Hokkaido, in the north, but rapidly dropped that manoeuvre and settled on establishing a beachhead feed mill at Shibushi, Kagoshima

Prefecture, on the southern island of Kyushu. There it met local resistance from the prefectural assembly.

At that point, Cargill turned to the US government to make 'a representation' to the Japanese government to allow it to build the plant under the required ruling by the Japanese Customs Office that the company could import feed ingredients without paying import duties. It was not long before the Kagoshima Prefectural government, in 1986, decided to sell Cargill the land it required for the feed mill, overriding the fears of other feed millers and farmers who feared that this would lead to integrated livestock production under Cargill's control.

The official account is that Cargill got permission because the livestock industry in Kyushu was expanding at the time and it was judged that another mill would not cause harm to existing feed mills, but the Korean Feed Association provided me with what is probably a more accurate account. They say that when Cargill tried to get into feed milling in Japan, the government would only permit them to buy an existing plant, not build a new one. But when Cargill tried to buy a plant, all the mills in Japan agreed not to sell. Then the US government intervened on Cargill's behalf and the Japanese government finally relented, giving Cargill permission to build a new plant. The trade-off was that unlike other importers of feedstuffs, Cargill was required to pay duty on what it imported. Whether they actually paid it or not is another question.

A candid insider's account of the whole process was provided by a news story in 1992 about Juels Carlson, president of Cargill North Asia Ltd from 1985 to 1990. Apparently Carlson was sent to Tokyo with the task of 'breaking through the investment barriers the Japanese government and agricultural industry had established to prevent internal competition'. According to the story, Carlson was successful in getting permission for Cargill to build its feed mill despite the opposition of that country's powerful agricultural industry.

> As Carlson explains it, Cargill had right, politics and common sense on its side. His strategy, in simple terms, was to refuse to take no for an answer, even the many versions of yes in Japanese that, in effect, mean no ... Cargill v.p. Wm. Pearce said Carlson was able to enlist the support of key Japanese politicians at the local level and in the Diet, the Japanese parliament. Eventually, the US trade representative in Japan weighed in on Cargill's

behalf, and the license was granted. In all, it took about five years and most of Carlson's time in Tokyo.[167]

In 1989 the requirement for ministerial authorization to import feed ingredients was lifted, but Cargill did not expand, preferring, apparently, to sell ingredients to other companies. On the other hand, perhaps the *zaibatsu* had forcefully indicated to Cargill the position it was to occupy. Cargill Japan's feed division is now just another feed company, with the exception of its function as an intelligence unit for its corporate parent, according to key industry people, who figure that Cargill does a lot of business with the Japanese trading companies.

On the other hand, what Cargill discovered, I was told in the Ministry of Agriculture and elsewhere, is that Japanese farmers are not just like those in North America. They are very demanding and expect their feed supplier to meet very precise specifications, whereas in North America feed is feed, and individual farmers add their own supplements. Japanese farmers, I was told by the Feed Trade Association, are also conservative and do not like switching suppliers, from whom they can get credit. So Cargill has found it preferable to import in volume and supply other millers in bulk, letting them have the headaches of dealing with the individual farmers.

Having tried seed and feed, it is hardly surprising that Cargill also tried fertilizer, where it ran into familiar opposition. When it proposed to build a fertilizer plant in Kumamoto Prefecture, the prefectural government rejected its proposal because of its possible impact on a local fertilizer plant. Cargill again had to go to southern Kyushu to establish its bulk fertilizer blending plant, in 1988, as a joint venture with Mitsui Chemical Company. All ingredients are imported.

Having less than stunning success with agricultural inputs, Cargill decided to move upscale (more 'value-added' in current jargon) to try to capture higher returns in one of its major product lines. Instead of just importing boxed beef from their plants in the US and selling it to Japanese processors and distributors, it decided to do the processing and distribution itself in Japan.

The first steps were to establish an office in Osaka in 1990 and make a deal with Daiei, Japan's largest supermarket operator, to carry Excel beef imported from the US and Canada under the brand name Kansas Beef. The third step, announced in early 1991, was to build a beef 'further-processing' plant in Japan. Consistent with its policy of holding a controlling interest in any joint venture, Cargill North Asia Ltd held two-thirds and Showa Sangyo Co one-third of the new company, the first foreign-owned meat processing plant in Japan. Showa Sangyo's contribution to the partnership was in the form of a newly built seven-story refrigerated warehouse in Funabashi, just east of Tokyo. Cargill's contribution was a portion-control processing plant built alongside. Together they were to build up their distribution network. Cargill expected Japan's imported beef market to increase steadily and it looked forward to controlling 30 per cent of the market. Barely two and a half years later Cargill halted its Japanese subsidiary's beef-processing and distribution operation and sold the processing plant to Nippon Meat Packers Inc., one of Japan's biggest meat packers, at a loss, it is said, of $10 million. (Cargill has had a joint venture with Nippon Meat to produce broiler chickens in Thailand and market them in Japan since 1989. Cargill runs the plant and Nippon sells the chicken.) Industry people said that a major problem was Cargill's manager, who had no understanding of the Japanese system and seemed to think that what worked in the US could simply be duplicated in Japan.

Cargill's long-term persistence paid off in 1995 when Cargill Japan Ltd became the first foreign firm to be granted the right to sell wheat, barley and rice directly to the Japanese Food Agency by the Japanese government. In other words, to be recognized as a 'Japanese' company. A year later Cargill decided to sell its animal feed-processing factory in Kagoshima Prefecture, which had been its initial 'beachhead' in Japan, to Chubu Shiryo for $11.5 million because increasing market share was too difficult. Even though it priced its products 20 per cent below prevailing levels, the domestic distribution system proved an insurmountable obstacle. The focus of Asian operations was to shift to China and south-east Asia.[168]

In 1998 Cargill Japan assumed management of Toshoku, a Japanese agrifood company that had filed for bankruptcy, with debts of $5.03 billion. Cargill was to serve initially as a 'sponsor' in accordance with Japanese law, intending to assume full ownership of Toshoku by spring 2000. Under Japan's Corporate Rehabilitation Law, a bankrupt company needs a sponsor to manage the company's reorganization. Once the courts and creditors had approved a reorganization plan, Toshoku could become a Cargill subsidiary. 'We are honored to be the first foreign company to be approved as a sponsor in Japan', said Hideyo Suzuki, president of Cargill Japan. Toshoku, established in the 1940s, traded in both commodities and processed food. It has about 20 product lines, eight Japanese offices,16 overseas locations, a chain of 24 supermarkets, an apple-juice plant in Washington state, a chicken-breeding farm in Illinois and a sugar refinery in Japan. A Cargill spokeswoman said Toshoku was actually quite similar to Cargill Japan. A seach of Cargill's websites turned up one for Toshoku in Japanese, so apparently it is now operating as an integral part of the Cargill empire.

China

Cargill's trade with China started with China's opening in 1972. Cargill sells grain, oilseeds and their products, cotton, fertilizers, malt, animal feed, cocoa, fruit juice, lecithin and soy protein, meat, salt, steel and other commodities to China, while buying Chinese commodities such as cotton, steel and corn for export. It started operations in China in1988 with an oilseeds-crushing plant.

Mike Hsu, manager of Cargill's grain department in Hong Kong, told me in 1994 that Hong Kong was Cargill's China office, at least until they opened an office on mainland China in the near future. An important component of the Hong Kong office, according to Hsu, was CIS. Hsu explained to me that Chinese people love to speculate, much more than western people, so Cargill thought that eventually, when the Chinese are rich and once they understand how to trade on the futures market, the Chinese would be very big speculators or hedgers. Cargill was getting ready, with the hopes of being able to be one of their brokers. 'Once China gets rich, they are going to be very big speculators. They don't have money, but they love to have money, to risk their lives', said Hsu. Cargill also had an investment team in Hong Kong which was responsible for planning future projects for Cargill in China. 'Our major focus will be on selling to China, not buying from China. It is so huge that eventually they will have to buy.' However, Hsu suggested, to be successful in China, western countries will have to change their trading strategies and companies will have to overcome the barriers of politics and language that stand in the way of their doing business in China. So what if it takes $5 million or $10 million and five years? The question is, what can we learn in that time for that money? Hsu said most of the people in the Cargill Hong Kong office spoke Mandarin, the official language of China, as opposed to Cantonese, spoken in Hong Kong.

Hsu articulated a common and crucial theme in discussions about China: internal transportation in China was very poor, the railroads in terrible condition and the country very large. As a consequence, it was cheaper to export food from the north and import food into the south than to transport it overland from the agricultural regions of the north-east to the industrial regions of the south. Of course this means that the food entering the south could come from anywhere in the world, and that food exported from the north has to compete with it. In other words, under its current pattern of development, China is being very rapidly integrated into the global food system of companies like Cargill.

In a 1996 feature on the future of Hong Kong and 'local' businesses, the Minneapolis *Star Tribune* interviewed several Cargill executives, among them vice-president for public affairs Rob Johnson, who said that Hong Kong would inevitably become less important as a beachhead for capitalism in Asia as countries in the region, including China, increase their commitment to freer markets. Daniel Huber, president of Cargill Asia/Pacific, said Cargill hopes one day to get Chinese contracts to introduce modern transportation and storage methods. The Chinese won't just be buying from the rest of the world, said Huber. They need to find ways to sell the raw materials and commodities they produce, and Cargill hopes to be at the centre of many of those trades. There are a lot of raw materials to come out of Asia, said Huber, salt being the biggest one, but also cotton, cocoa and rice and rubber.[169]

In 1998 Cargill moved its head office for China region to Shanghai and by the end of 1999 it had 650 employees in eight sites in China and was doing about $1 billion a year in business with China, nearly three-quarters of it in the form of imports.

Cargill manages its information as carefully as it manages its businesses. It has long been highly secretive about its activities in China, and when it does provide information, it usually raises as many questions as it answers. For example, there is no date on the website item from which the following information was obtained, but one can guess that it was late 1999 from the fact that it cites the company's 1999 financial report and fiscal 1999 ended May 31, 1999:

> Cargill Investments (China) Ltd. is a wholly-owned investment company headquartered in Shanghai ... Cargill is considering future investments in corn milling, oilseed processing, feed production and other agricultural industries. Cargill is a partner

in the Sino-Foreign joint venture trading company, Dongling
Trading Corporation, based in Shanghai. Through Dongling,
Cargill provides products and services directly to Chinese end-
users. Dongling's agricultural division, for example, imports and
distributes products such as oilseeds, soybean meal, fruit juice concentrates and other food ingredients to customers throughout
China. Dongling Trading also exports such products as food beans
to international markets.[170]

Cargill opened its first bulk fertilizer blending facility in China in
1996. The new subsidiary, Tianjin Cargill Fertilizers Co. Ltd, supplies
agronomic information and services to corn, wheat, vegetable and
fruit farmers as well as fertilizer. In 2000 Cargill opened a second
fertilizer plant, Yantai Cargill Fertilizer, in north-eastern China. It
also joined with Yunnan Phosphate Fertilizer Factory to invest $15
million each in DAP (diamonium phosphate) production facilities
in Yunnan Province, with Cargill managing sales for the joint
venture. Yunnan has the largest deposits of phosphate in China with
4.2 billion tons of verified reserves around Dianchi Lake. Cargill
currently supplies China with about 1.2–1.5 million tons of quality
DAP annually. Cargill also has a joint venture with High Value Feed
Corn Company in Liaoning, but again, there is no further information available explaining what this might be.

However, *People Daily* has reported that in partnership with
Taiwan's largest food company, Presidents Group, Cargill is building
a $120 million soybean plant in Guangdong Province in south
China. The two companies, already collaborators in a number of
enterprises in Taiwan and elsewhere, are equal partners in the new
company, President Cargill Fodder Protein Co., which is to have an
annual processing capacity of 1 million tons of soybeans. It will be
Cargill's largest investment project in China. Cargill is the largest
purchaser of China's corn products, and its annual trade with
China's mainland has averaged $1 billion in recent years. Since the
Taiwan President Group entered the 'inland' market in 1991, it has
set up 45 companies. Guangdong has been its major investment
market due to its convenient location.[171]

In 2001 Cargill, together with Global Bio-Chem Technology
Group of Hong Kong, formed a new company as a vehicle for
investing in the manufacture and sale of sweeteners and bio-
fermentation products made from corn processing in China. The first
investment will be a 50–50 owned HFCS refinery to be built in

Shanghai. Global Bio-Chem will finance and manage the construction and operations of the Shanghai refinery, while Cargill will contribute design, engineering, quality assurance methods and procedures to support the plant's construction and operation. Global Bio-chem is described by Cargill as 'the key corn wet milling company in China'. Shortly after the announcement of this project, Global Bio-Chem announced that it was negotiating an investment of approximately $97 million to build an ethanol plant in China. There was no mention of Cargill, but it is likely that Cargill will be involved in the project.

Asia Pacific

It's a challenge to keep Cargill's management structures straight because they seem to be highly fluid, which, of course, is one reason for the company's success over the years. Cargill is neither structurally nor ideologically rigid. Cargill has operated out of Hong Kong, Shanghi and Singapore in a number of capacities. Hong Kong was their China office, now it is Shanghai. Singapore has been the office for Cargill Asia Pacific since 1974 or earlier. Cargill Asia Pacific appears to cover the countries to the south of China such as Thailand, Vietnam, Indonesia and the Philippines. In 1995, Cargill Asia Pacific was responsible for operations in 16 countries.

At that time Cargill said it planned to invest some $1.5 billion in Asia in the next ten years, mainly in grains and oilseeds, with the emphasis on oilseeds processing, animal feed and poultry. It already had poultry production in Sri Lanka, Indonesia and Thailand and palm-oil refining in Malaysia. The company planned to set up a palm-oil plantation and refinery in Indonesia and to expand its existing Malasian palm-oil refineries. In 1997, Cargill Siam opened a fertilizer blending plant in Thailand, 95 km south of Bangkok near the deep-water port of Siam.

Cargill's beachhead in Vietnam, as in so many other places, was a feed mill, followed two years later, in 1998, by the construction of a swine technology and extension facility designed to increase the use of high-quality feeds and improve swine production in Vietnam. Training sessions run by the US Grains Council are based on the 52-sow farrow-to-finish operation because Cargill thinks that size piggery is a realistic goal for Vietnamese producers, most of whom have two to five pigs in their household-type operations.

Whitney MacMillan, billed as 'Chairman Emeritus Cargill', made a policy speech at the dedication ceremonies for the feed mill in Vietnam in 1997:

> I was fortunate to attend the groundbreaking for Bien Hoa Feed Mill in 1996. I am doubly fortunate to be back today to see first hand just how much American/Vietnamese cooperation can accomplish in such a short time. Like many business leaders of Vietnam, Cargill is a strong advocate of free and open trade. A facility like this is ... a living example of Cargill's corporate vision of helping to raise the living standards of people throughout the world ... Traditionally, Cargill takes a conservative, long-term approach to making investments. We look to areas in which we can extend the basic skills we have developed in existing businesses to areas in which they are both needed and wanted. We take these basic businesses into new countries, first on a small scale as we have done here in Vietnam. The business may be seed research and production, or basic processing, or feed manufacturing, or others. We use this proven business base to learn about the market and the social, political and economic environment of the new country.[172]

Cargill Philippines Inc. began construction in 1998 of two feed mills, one in Baliuag, Bulacan, and the other in General Santos City, South Cotabato. The two facilities produce 300,000 metric tonnes of animal feed yearly. 'We are excited by the opportunity to leverage our global feed technology and expertise for this market, by which we can deliver increased value and productivity to the local producer', said Randy Seibel, general manager of the Animal Nutrition Division of Cargill Philippines. The simultaneous construction of two plants was Cargill's largest capital investment in the Philippines up to that time. Cargill already employed 320 people in its Philippine businesses, which included copra processing, hybrid-seed-corn and protein-meal distribution.

In 1995, the majority of the 6,000 corn farmers in the southern Philippine agricultural town of Banga found their fields ravaged by a disease called 'stalk rot'. Much of the yellow corn they harvested contained only a few kernels and the yield was far below expectations. The culprit, they said, was the product of modern agricultural technology – hybrid seeds developed by Cargill and given free to them by the agriculture department, which promised farmers they could double their yield. The yellow corn seeds were part of a five-year, $2.8 billion programme designed to make the Philippines self-sufficient in grain. Developed by Cargill in other countries, among them Thailand, the seeds were bought under the Philippines' Grains Production Enhancement Programme – a package of infrastructure, credit and agricultural subsidies given to farmers as an incentive to raise productivity by using high-yield hybrid seed varieties. The farmers in Banga said the Cargill seeds could not stand the wet and humid climate of South Cotabato and required fertilizers and pesticides. These imported inputs raised farmers' production costs and increased dependence on TNCs.[173]

17 Seeds

The story of Cargill's seed business tells a lot about corporate strategy, in spite of the paucity of information, just as a detailed examination of Cargill's activities in any other commodity or geography does.

In 1994 the industry journal *Seed World* put Cargill Hybrid Seeds in the category of 'industry giants'. Six years later it was out of seeds altogether, although the buyer of its global seeds business was Monsanto and at the time of the sale the companies also formed a joint venture called Renessen (more on that shortly). The seeds business is not a public-attention grabber, and this alone would provide adequate explanation as to why Cargill frequently used seed, as the Trojans used their famous horse, to establish a beachhead in a new geography. Only a carefully targeted consituency of farmers need ever know anything about Cargill's activities in the seed business and their loyalty to their seed supplier could supply Cargill with frontline troops for further advances in alien territory.

Although the seed used in crop production is not a high-volume bulk commodity, there is no money to be made out of transporting or storing it, it is not traded on the futures markets (so no money can be made speculating in it), and there is no processing required to speak of; there is, nevertheless, good money to be made in hybrid seed, and more recently, patented and genetically engineered seed, all of which sell for five times as much, or more, as open-pollinated seed. The biggest return, however, comes from the addiction and dependency that can be created by means of hybrid seed: addiction to chemical fertilizers (a Cargill specialty) and agro-toxins (which Cargill sells but does not manufacture); and dependency on the seed company for new seed every season. ('Hybrid', as used in the seed business since the 1930s, refers to seed that is produced by crossing two inbred parent lines. The first-generation progeny benefit from 'hybrid vigour' but, thanks to the inbreeding, the progeny do not 'breed true'. That is, the genetic makeup of the seed from the crop falls apart if it is used as seed for next year's crop. The pay-off for the seed company is that the farmer has to return each year to the dealer for new first-generation hybrid seed.)

Will Cargill apparently started experimenting with seed breeding in the last century, and his company has been in the seed business ever since 1907. When 'modern' hybrid corn was invented in the mid-1930s, Cargill was not long getting into it, marketing hybrid seed under the Crystal Brand name and utilizing the breeding stock developed originally at the public (land grant) universities of Minnesota and Wisconsin.[174] Years later, in 1953, the company was charged with adulteration of seed, that is, selling seeds that were not what the label said. It had been adulterated with inferior seed, weed seed or even simply dirt. Cargill settled out of court when it became apparent that at least some of the charges were true. It also left the seed business except for its hybrid corn operation because its reputation had been so badly tarnished that the company recognized that the business would be bad for some time.[175]

Hybrid seed is attractive as a means of establishing beachheads in new territories because it requires virtually no capital investment. Practically all the company has to do is send a salesman with a few bags of seed, an airplane ticket, and enough money to buy a motorbike. Cargill strategist Jim Wilson has described Argentina in the 1960s as Cargill's 'first major beachhead' where the product line used was hybrid corn seed. Tanzania and Turkey are also good expressions of this strategy. In Tanzania the manager works with a staff of 24, most of whom are involved in seed production. Four or five of the staff 'bounce around the country on dirt bikes setting up a dealer network' and selling and delivering seed in small quantities of 1–10 kilos. The managers work with 'contract seed growers who run much bigger farms than most of their customers'.[176]

In 1991 Cargill expanded the corn and sunflower seed business that it had established in Turkey in 1987 by building a $1 million seed-conditioning facility in the south of the country. Until the mid-1980s, according to Cargill, Turkish farmers planted traditional open-pollinated varieties. Then changes in government policy encouraged companies like Cargill to introduce hybrid seed. The addition of their new Turkish plant brought the number of countries in which Cargill had hybrid seed operations to 27.

Cargill went along – for a while – with the mid-1990s fad of acquiring seed companies with an eye to a vertically integrated system, acquiring, among others, Goertzen Seed Co. in 1994 and Vineyard Seeds the next year. It even got into distributing insect-resistant Bt* seed corn under the NatureGuard brand of Mycogen

---

* Bt refers to the toxin removed from the soil bacterium *Bacillus thuringien-sis* and inserted into the genome of maize and cotton.

Corp. It also began to fund the work of Mogan International of the Netherlands on fungus-resistant canola and sunflower varieties, and on the Binary Vector System, a patented method for the genetic transformation of crop plants. Cargill planned to market its canola and sunflower varieties containing Mogen's fungus-resistance gene worldwide.

Then in 1998 Cargill abruptly sold its international seed operations in Central and Latin America, Europe, Asia and Africa to Monsanto for $1.4 billion. The acquisition included seed research, production and testing facilities in 24 countries and sales and distribution operations in 51 countries, but did not include Cargill's seed operations in the US and Canada or Cargill Agricultural Merchants in the UK. At the time, Monsanto was buying up every seed company it could get its hands on anywhere in the world. As a consequence, by 2000 Monsanto had a debt of $8.1 billion.

Three months after that deal, Cargill announced the sale of Cargill Hybrid Seeds North America to Hoechst Schering AgrEvo GmbH (AgrEvo) for $650 million, but the sale collapsed early in 1999, as a consequence of a lawsuit brought against Cargill by Pioneer Hi-Bred International Inc. charging that Cargill wrongly obtained some Pioneer genetic material for use in its own seed research and development programme. After investigating the situation, Cargill reported that it had 'uncovered a previously unknown problem with genetic material that had been introduced into our breeding programme by a former Pioneer employee'.[177] In May, 2000, Cargill and Pioneer announced that they had settled the lawsuit. Under the settlement, Cargill agreed to destroy misappropriated material in its corn-breeding programme, not to engage in the practice of isolating parent seed from bags of Pioneer's hybrid seed corn and to pay Pioneer $100 million for past damages. In announcing the settlement, Pioneer's president, Jerry Chicoine, once again underlined the high degree of cooperation and good will that exists between supposedly competitive TNCs:

We have made huge investments in seed research and development and take our intellectual property rights very seriously. Fortunately, Cargill also took those concerns seriously and to its credit did a thorough job of investigating and eradicating problem areas it found in its seed business. That determination to make things right made this settlement possible.[178]

Cargill executive vice-president Fritz Corrigan played the part of innocent victim eager to appear morally upright:

> This has been a painful period for Cargill; we were shocked that our investigation into Pioneer's allegations revealed that our seed business hadn't always lived up to our high ethical standards, but we have learned from this experience, we have honored our commitment to make things right and we have emerged with a solidly respectful and stronger relationship with Pioneer and DuPont.[179]

Without explanation, Brian Hill, head of Cargill's seed operations, left the company suddenly in June, 1999 after being with the company for 31 years.

In the meantime, having turned away an earlier bid from Monsanto, in August, 1997, Pioneer sold 20 per cent of itself to E.I. DuPont de Nemours & Co. for $1.7 billion. In March, 1999, DuPont bought the rest of Pioneer Hi-Bred International for $7.7 billion.

The intimacy of corporate affairs is further revealed by one of the consequences of DuPont's acquisition of Pioneer, an alleged conflict with a non-compete clause signed when DuPont sold its Inter-mountain Canola business to Cargill in 1994. To avoid litigation, Cargill released DuPont from the non-compete commitment in return for some technology and other non-monetary considerations.

When the legal tangle was finally resolved, AgrEvo had lost interest, but Cargill was able to sell the remaining assets of Cargill Hybrid Seeds to Mycogen Seeds, a subsidiary of Dow Chemical Company, in September 2000, for an undisclosed price. Dow integrated these assets into Mycogen Seeds and created a new seed organization within its Dow AgroSciences subsidiary. The purchase included all seed-research, production and distribution facilities of Cargill Hybrid Seeds in the US and Canada, except for Cargill's Inter-Mountain Canola, Goertzen Seed Research and the Western Canadian seed distribution business. There has been no public explanation of these exclusions.

When Cargill sold its global seeds business to Monsanto in 1998, it formed, at the same time, a worldwide joint venture with Monsanto to create and market new products 'enhanced through biotechnology' for the grain-processing and animal-feed markets. According to the Cargill press release:

The 50–50 joint venture draws on Monsanto's capabilities in genomics, biotechnology and seeds and on Cargill's global agricultural input, processing and marketing infrastructure to develop and market new products with traits aimed at improving the processing efficiencies and animal nutrition qualities of major crops.

The name of the new venture is Renessen, which, as Cargill explained, is derived from the word Renaissance, 'a period of rapidly accelerating knowledge that signalled the beginning of a new age'.[180] At the end of 2001, on Cargill's website under 'joint ventures, Rennesen', one can find the standard 'single gene' line about genetic engineering being a 'more controlled' form of traditional plant and animal breeding:

Agricultural biotechnology is simply the application of scientific discoveries to the age-old human quest to produce more and better food for our families and communities ... Modern agricultural biotechnology allows researchers to insert a single gene into a plant to give it a specific trait. It is a more controlled method of producing plants with certain desired traits.

The information trail on Renessen goes no further, but another indication of the 'downstream' direction Cargill is moving in is found in the name Cargill Health and Food Technologies which Cargill gave the new division that brought together its nutraceuticals and specialty food ingredients businesses in 2001. The division is described as 'a leading developer, processor and marketer of science-based, health-promoting ingredients for food and dietary supplements industries worldwide.'[181]

Cargill Seeds India

Back in 1983 Cargill decided to establish a beachhead in India via seed, but it was not until 1992 that it was able to actually implement this strategy, using hybrid corn and sunflower as the advance troops. As in other areas where Cargill has pursued this strategy, its long-term goals far exceed the simple marketing of seed. India, like Argentina and Turkey, has the potential to become a major grain- and oilseeds-growing region for Cargill's global system of processing and marketing. The northern states of Rajasthan and Punjab could, with the benefit of irrigation, become global sources of grains, while the south-central region, including the states of Karnataka and Maharashtra, could become major sources of corn and oilseeds. These regions now produce a wide variety of foods for the people of India, and women play a vital role in food production, from selecting and conserving seed to caring for and harvesting the crop. What happens to women and what the people would eat under the Cargill regime is another question altogether.

To understand Cargill's moves in India, however, it is necessary to understand something of recent Indian policy changes regarding foreign enterprise, trade, and ownership rights, whether of seed, land or business itself.

In July 1991, the government of India fell into the arms of the World Bank/IMF and ushered in a new economic and industrial policy with a marked devaluation of the rupee. At the time the economy was in a crisis of sorts: there had been a steady increase in the fiscal deficit for more than a decade and the country was on the verge of default on external payments. The devaluation of the rupee was quickly followed by massive external loans and an IMF-guided policy of economic changes through structural adjustment.

By 1995 the industrial sector appeared to be in recession, partly due to the collapse of numerous small- and medium-scale enterprises as the result of liberalized imports. Nearly half of the Indian corporate sector was said to be 'sick':

> Indian society is getting more and more expenditure-oriented and heading for consumerism with liberalised imports favouring ostentatious consumption of the rich and privileged ... Multinational financial institutions as well as transnational corporations are gaining the upper hand while indigenous financial institutions and the Central and state governments are playing a docile and subservient role of meekly accepting the consequences of foreign economic aggression.[182]

Under the new economic policy, foreign equity in Indian companies could be increased from 40–50 per cent without government permission or even prior consent of the shareholders. Further liberalization would make 100 per cent foreign ownership possible while providing for the transmission of profits and royalties abroad. For example, Indian-owned Tata Oil Mills Company (Tomco) merged with Hindustan Lever, while Godrej Soaps merged with Procter &

Gamble. PepsiCola raised its equity position to 51 per cent of the Voltas company.

India's 'New Policy on Seed Development' in 1988 encouraged increased collaboration between domestic and foreign companies in order to increase the import of technology and genetic material and to encourage private seed companies with the objective of providing Indian farmers 'with the best seeds and planting materials available in the world to increase productivity'.[183] The new policy reduced the duty on imported seeds from 95 per cent to 15 per cent, but stipulated that importers had to submit adequate quantities of seed to the Indian Council of Agricultural Research (ICAR) to be used for trials at 12 to 15 different locations and to be stored in genebanks. It was obviously no mere coincidence that when approval was finally received from the Indian government in 1988, Cargill began to implement its 1983 decision to enter the seed business in India by forming a joint venture with Tedco, a Tata company (51 per cent and 49 per cent respectively).

John Hamilton, managing director of Cargill Seeds India, set up office in Bangalore and 'research' operations began the following year, with the marketing of hybrid sunflower and corn seeds starting in late 1992, according to Hamilton. The 'research' Cargill carried out in India consisted of limited seed trials and the selection of varieties of imported germplasm it deemed to be suitable. A company press release trumpeted: 'All Cargill hybrids sold in India are derived from imported germplasm as part of the Cargill strategy of providing the best genetics of the world to the Indian farmer.'[184]

Such a statement expresses a deeply colonial attitude as well as a highly unscientific approach to genetics. Germplasm is the expression of the relationship between an organism and its environment at a particular moment. In practice, Cargill, along with the entire biotech industry, acts as if the organism exists apart from any particular space or time. Not only is this bad science, it also expresses contempt for indigenous knowledge in general and traditional plant selection and genetic conservation in particular.

The emergence of Cargill in India as a seed company, coupled with the push to conclude the Uruguay Round of the GATT negotiations (with its strong emphasis on intellectual property rights, including the patenting of seeds), ignited a campaign againt the GATT and Cargill in December, 1992:

Over 500 farmers, belonging to Karnataka Rajya Ryota Sangha
(KRRS) [the farmers' movement in Karnataka], stormed the office
of the American multi-national, Cargill Seeds India Private-
Limited, on the third floor of a building on St. Mark's road and
threw out all the papers and files onto the road and burnt them
here today ... They also threw out the seed-samples kept in the
office.[185]

While it was reported in the press that no company official was
attacked, in keeping with the KRRS policy of non-violence, and the
office personnel never claimed that they were harmed in any way,
Hamilton commented to the effect that the KRRS needed to be
reined in, and said to the public: 'Let them do anything within the
law, but let them not smash Cargill.'

Undeterred, in early 1993 Cargill started to build a seed-processing
factory on a 13-hectare site at Bellary, 300 km north of Bangalore.
The facilities were to include an administration and seed-technology
training centre 'to develop modern agriculture', according to
Hamilton. The plant was scheduled to begin production in October,
1993, but early in the morning of July 13, the farmers of the KRRS
gathered at the site and with poles and their bare hands demolished
the partially completed facility. In January, 1994, I saw for myself
that Cargill was rebuilding, but progress was delayed because it first
constructed a fortress around the premises, complete with high
granite walls and guard towers, to protect Cargill from the farmers.

The leaders of the KRRS took me to villages and towns to talk to
farmers and seed dealers and everywhere the story was the same:
Cargill's hybrid sunflower seed produced only a fraction of the
advertised yield, no matter how strictly the suggestions for growing
were followed, and no matter how costly the fertilizers and
chemicals. The Rallis seed dealers, the distributors Cargill had
contracted for distribution, decided they would be better off not
selling Cargill sunflower seed.

Following the attack by the KRRS on the Cargill facilities at Bellary,
police began to provide full security to all Cargill structures. But as
the bill rose, Cargill refused to pay it, in spite of Hamilton's
statements that Cargill did not wish to impose a burden on the
police. 'But now Hamilton has gone back on his words and is saying
that providing security is the State's job and that the company would
not make its own security arrangements. The bill has bounced and

the police are hapless as the multinational is applying pressure tactics through the US embassy.'[186]

Mangala Rai, then of ICAR, told me in an interview that India is a big potential market for sunflower seed. For example, five years earlier no sunflowers were grown in Punjab, but by 1993 there were 100,000 hectares. Rai said the varieties that Cargill was making available in India were not the best on the market. Cargill may have better varieties, Rai explained, but they did not want to bring them here because India did not have varietal protection. Rai said Cargill simply does not listen to Indian advice. What they were selling was not what ICAR recommended on the basis of its trials.[187]

M.D. Nanjundaswamy, a Karnataka state legislator and the leader of the KRRS said:

> It took us 12 long months to bring the farmers to this point of resistance on the seeds issue and to organize and educate them on intellectual property rights and the Dunkel Draft. By and large the Indian farmers are illiterate, but they could understand this issue because they live with the plants every day and they live with the seeds every day. That is one of the reasons they could understand it. Because of their relationship with plants and seeds they could even understand genetics much faster than the Indian intellectuals. Since the mid-1960s, more than 25 years of direct experience of the Green Revolution and its culture helped in making the farmers understand how GATT is working out to become the second chapter in the enslavement of farmers, following the so-called Green Revolution.[188]

Cargill's Hamilton was quite specific about Cargill's intentions in India. 'The possibility of shifting production in India is significant and we know that where we pick up farmers who, shall we say, are focused on development, and we give them good genetic material and we give them good training and we hold their hand through the growing system – you give these farmers two seasons, and boy, their productivity changes by factors of like 100 per cent.'

The Cargill–Hamilton vision for Indian called for large-scale production of corn and oilseeds – sunflowers and rapeseed/canola – in the south of India, and small grains in the north, which was already much more mechanized than the south. 'I think India can become a wheat exporter. If we can harness population growth in India, then wheat exports are a real possibility. Industrial use of corn is on the increase, and if we grew more corn there would be more industrial use.'

In discussing the very touchy issue of intellectual property rights, Hamilton made Cargill's strategy quite clear: proceed without them while lobbying hard for them. On the one hand Hamilton says: 'What India does about intellectual property rights won't make any difference to us. We made a conscious decision to operate in India many years ago', while in almost the next sentence he told me: 'Through the Seed Association of India, we have been lobbying the government already – well, I've been here for six years and we have been lobbying for six years.' As for other agricultural ventures, Hamilton said that Cargill was already importing a significant volume of fertilizer into India.

Hamilton, being something of a one-man show, did some of his own PR work. He gave me a copy of *Our Link*, 'a Cargill Seeds India Publication, October, 1993, published by John Hamilton on behalf of Cargill Seeds India Pvt Ltd for Private Circulation Only'. This was simply an attempt to discredit the farmers criticizing Cargill:

> A group associated with KRRS has been involved in a campaign of misinformation and two specific instances of violence against Cargill. The allegations levelled by the group and its leader have already been contested and refuted by intellectuals, farmers, the media and Cargill itself.

Judging by the newspaper clippings and Hamilton's failure to identify his supporters, it was clear that Cargill had few friends to come to its aid.

Cargill Seed in China

Two brief accounts of commercial relations regarding seed between Cargill and the government of China give some insight into the importance attached by companies like Cargill to secrecy and the ownership of information of all sorts, including genetic. I have been unable to track down any information beyond two reports summarized here, in spite of my intense curiosity.

Some years ago China developed a commercially attractive rice hybrid, capable of increasing harvests by up to 25 per cent. In 1981 China granted exclusive rights for development, production and marketing of this seed in specified countries to Cargill Seeds and Ring

Around Products Inc., a subsidiary of Occidental Petroleum. The hybrid rice variety apparently covered more than a third of China's 33 million hectares of rice paddy in the mid-1980s and was a so-called 'male-sterile' line of rice that would not self-seed and thus could be easily crossed with another variety. However, an agreement between the government of China and the two companies forbids the sharing of information and materials concerning hybrid rice with other governments or with the International Rice Research Institute (IRRI), which is backed by Rockefeller and Ford money. The IRRI found out about the agreement in 1987 when it discovered that something forbade the sharing of information and materials concerning hybrid rice with other governments or with the IRRI.[189]

18   Salt

Salt attracts little attention, being everywhere and nowhere. It is so cheap, it seems, that we probably never really notice the price or read the label, much less wonder where the few cents really goes. But as with all bulk commodities, it's all a matter of volume. If you can handle enough of it, you don't need to make much on this pound or that kilo, but like grains and oilseeds, salt is an essential. This is the type of business Cargill knows and likes, so it should not really come as a surprise that Cargill has been involved in the salt business for a long time. In fact, Cargill estimates that it has 10 per cent of the world salt business, including both food grade and industrial/road salt:

> Cargill entered the salt industry in 1955, when it purchased 750 tons of rock salt from the Jefferson Island Salt Company to fill an empty grain barge. Cargill managed to sell the salt, using the barge as a warehouse. In 1960, Cargill Salt acquired the mineral rights to Belle Isle, La., and began work on its first rock salt mine. By December 1962, mining had begun and by 1967, production had reached 800,000 tons. Since then, the company has expanded its reach into every major salt market.[190]

Salt making around San Francisco Bay dates back to the Indians who once tended about 800 hectares in the South Bay. In the 1850s, small works produced salt for gold mining. Leslie Salt Co. came to the bay in the 1930s and in 1977 the US Federal government condemned 6,140 hectares of Leslie Salt's land, including 4,800 hectares of diked salt ponds, for the national refuge which had been established five years earlier. As part of the $7.6 million condemnation deal, Leslie retained the rights to extract salt in perpetuity from the ponds. In 1978, Cargill bought out the Leslie holdings, including the right to extract salt from those salt ponds in the refuge.[191] At the same time Cargill acquired Leslie's solar salt facility at Port Hedland, Australia. Cargill took over the rest of Leslie Salt Co. in 1991.

In 1995 Cargill de Venezuela acquired a 70 per cent interest in Productora del Sal (Produsal), a Venezuelan company which had started construction of a solar salt facility adjacent to Lake Maracaibo in Zulia state. The facility began producing in 1999, and when the plant reaches full capacity in 2003, it should be producing 800,000 tonnes of industrial quality salt per year and exporting 300,000 tonnes of it annually (with Europe indicated as the major market). Venezuela uses 500,000 tonnes of industrial salt per year and the production from the Cargill facility will eliminate the need to import salt for the production of chlorine used primarily in the production of polyvinyl choloride (PVC) plastic.[192] (More on this below.)

In the mid-1990s just four companies – Morton, Akzo-Nobel of the Netherlands, North American Salt Company and Cargill – dominated the salt business in North America. North American Salt, a subsidiary of Harris Chemical Group of Kansas, was formed in 1990 out of Sifto Salt, American Salt, Carey Salt and Great Salt Lake Mineral companies. (Carey had acquired Sifto from Domtar Inc. of Montreal in 1989.) It had mines in Cote Blanche, Louisiana and Goderich, Ontario. The Goderich mine is probably the largest salt mine in North America.

Although they are supposed to be competitors, Morton has a packaging facility next to Cargill's salt production facility in SF Bay that packages Cargill's salt under the Morton label. (Morton was purchased by specialty chemical company Rohm & Haas for $4.6 billion in cash and stock in 1999.) Akzo-Nobel operated rock salt (for road de-icing) mines in Louisiana, Ohio and New York state and solar salt plants in Utah and in the Netherlands Antilles. Akzo also operated a large refinery, producing food-grade salt as well as salt for the chemical and pharmaceutical industries, on Lake Seneca near Watkins Glen, New York. At this mine, salt is extracted by pumping fresh water from the lake through bored wells into the rock salt deposits about 600 m below the surface. The dissolved salt returns to the surface as brine for refining through an evaporation process.

A few kilometres from the Akzo plant, and right in the middle of the town of Watkins Glen, is a similar refinery owned by Cargill, acquired in 1978 and expanded in 1994. Not far to the east of Watkins Glen in Lansing, New York, just up the shore from Ithaca on Lake Cayuga, Cargill operates its only underground salt mine in the US, which it acquired in 1969. Here rock salt is mined, much like coal, from a depth of 600–700 m. The mine produces about 1 million tons of rock salt per year for road de-icing throughout the north-east US. The salt deposits which these three mines are exploiting were formed some 300 million years ago by evaporating sea water pools and form a large basin underlying much of Pennsylvania and parts of Ohio, New York and Ontario.

In 1997 Cargill bought the North American salt operations of Akzo-Nobel. Terms were not disclosed. The purchase included all Akzo-Nobel locations in North America and on the Caribbean island of Bonaire, except the salt mine and distribution centre in Retsof, NY. Akzo-Nobel's Diamond Crystal and other consumer brands were included in the transaction, as well as solar salt terminals in Port Newark, New Jersey, and Cape Canaveral, Florida. Cargill subsequently reported its worldwide salt-production capacity as approximately 14.7 million tons annually, including 7.2 million tons of rock salt, 1.8 million tons of evaporated salt and 5.7 million tons of solar salt from 30 production facilities located in the US, Venezuela, Australia and the Caribbean.

Cargill sold its liquid calcium chloride business to Tetra Technologies in 1998. The transaction included 320 hectares of solar salt production land in New Jersey, and 460 placer claims on US Bureau of Land Management property, all leaseable mineral rights, and a solar salt plant that uses solution mining for its brine source. Liquid calcium chloride products from the facility are used by a variety of food-processing and industrial customers.

Cargill's Salt Division opened a facility in the Port of Tampa in 2000 that combined the warehousing and marketing of evaporated salt produced at its plants in Michigan, New York and Ohio and the unloading of vessels of imported solar salt which would be made into pellets or other packaged products. In 2001 Cargill sold its Australian solar salt plant at Port Hedlund to Rio Tinto Group for $95 million, with a provision for increased payments, depending upon how well the business performs.[193]

Cargill's shuffling around in the salt business has obviously been complex, and at times rough, but it is an essential commodity and it must be profitable enough to be attractive, given the hassles and roadblocks the company has had to contend with. Stories about three solar salt sites – India, California and Venezuela – illustrate this.

India

While Cargill Seeds India was trying to establish its beachhead to the south, Cargill Salt was attempting to stage an invasion in the northwest on a more literal beach. It had its eye, in fact, on a potential global salt source in a major Indian port area, and while there might eventually be a market for the salt, it would appear in the context of Cargill's overall strategy that the real prize would be the port itself. The attractiveness of this prize encouraged Cargill to make some very aggressive manoeuvres. The ensuing story is one of the best documented of Cargill's efforts to get what it wants.

'Encouraged by India's new economic liberalisation, Cargill Southeast Asia obtained approval from the government's Foreign Investment Promotion Board [in August 1992] to set up a 100 per cent export-oriented unit to produce one million tonnes of high-quality sun-dried or solar industrial salt a year' [in Kandla Port, Gujarat State], reported the *Financial Times* in mid-1993.[194] Even though it was already producing 5 million tonnes of salt a year at its plants in Western Australia and California, Cargill was seeking new production sites because these sources were not expected to meet future demand.

'The island of Satsaida Bet, created by a system of inter-connected creeks, is perfect for the setting up of salt pans, but the silting could cause major technical and ecological problems', commented the *Financial Times*. The Island of Satsaida Bet is in the Kandla Port District at the head of the Bay of Kutch in the north-west corner of India. In addition to the salt manufacturing facility, Cargill was given permission to build a $25 million jetty capable of loading 10,000 tonnes of salt a day, compared with loading capacity of 1,000 to 2,000 tonnes a day at other Gujarat docks.

Cargill had originally planned to produce salt in collaboration with an Indian company, Adani Export Co. of Ahmedabad, and had applied to build a fast-loading jetty near the twin ports of Mandvi and Mundra, about 50 km west from Kandla. That deal was reported at the time as having 'fallen through', though perhaps the real reason it fell through was that, in the face of the moves of the Indian Federal government to 'liberalize' the economy and privatize ports, Cargill no longer felt the need for an Indian partner. After that Cargill pushed ahead with its private project in Kandla Port.

With Federal government permission in hand, Cargill asked Kandla Port Trust (KPT), which manages Kandla Port, to release 4,000 hectares of land on Satsaida Island. This request was turned down and under pressure from the central government, the KPT trustees met again and all those present agreed that Cargill could not be given permission. The representative of the Ministry of Defence also agreed because the island in question is considered a strategic site, being very close to the border with Pakistan.

The background to the KPT refusal is as follows:

1) The Government of India had allocated a total of 300,000 hectares of land throughout the country for salt production, but only half of this was actually being utilized to produce salt. Although there was a certain demand for Indian salt in the international market, this demand was not being met due to a lack of adequate infrastructure, particularly transport vessels and loading facilities. The need to improve infrastructure was recognized, but there was no need for greater production of salt.

2) Many research organizations advised KPT not to use the land Cargill was after for ecological reasons. Land on which salt is to be produced should be at least 7 m above sea level, but Satsaida Island is only 6.5 m or less above sea level. If Cargill were given the land to produce salt, its earnings could be in the hundreds of thousands of rupees, but the public costs of removing silt from the harbour would be in the millions because Kandla Port is designed to work with the tides.

3) The proposal would adversely affect the local mangroves, creating an ecological imbalance.

4) There would be a good chance that the 25,000 people currently involved in salt production and other activities in the port area would lose their livelihoods because of the project.

Apparently feeling that it had the political muscle to override ecological and social concerns, Cargill went again to the central government and this time KPT was directed to enter a caveat in the civil court so that no one else could go to court to block Cargill's application. However, one of the associations of small salt producers of Kutch had already gone to court and obtained a stay on any decision being taken by this extraordinary meeting of the KPT. This pre-empted the move requested by central government and the meeting ended with no decision made.

Local opposition to the Cargill project took the form of a protest march beginning in surrounding villages and timed to arrive at Kandla Port on May 17, 1993, the anniversary of Mahatma Gandhi's salt march to the sea in the same state more than 50 years earlier. Gandhi's march was one of the events that brought about India's independence from British rule and is a powerful image in the political landscape of the region. The protest march was organized in the Gandhian tradition of a *satyagraha*, a form of political action developed by Gandhi rooted in Indian culture and religion. The word *satyagraha* refers to the active force of love; it is an act undertaken to overcome evil through the active power of love. An aspect of *satyagraha* is a willingness to take upon oneself the violence and suffering caused by the evil and in this way bring it to an end. Thus the protest march against Cargill was much more than the term 'protest march' would connote in the industrialized West.

In September, Cargill made a tactical retreat from salt in India, saying it was no longer interested in building an export-oriented salt works in the KPT area and describing it as a business decision. Cargill said that the worldwide recession, and especially the slowdown of the Japanese economy, meant that there was no longer a need to build a world-scale salt works in and for the Asian-Pacific area at that time. The company also said that political opposition had played no role in Cargill's withdrawal from the salt project.

Ashim Roy, a union organizer in the region, thought that Cargill came for the jetty and the port itself. 'Of course they also wanted a monopoly on salt for export, but actually what they wanted was a bulk handling port. Kandla is 500 km closer than Bombay to the grain heartland of India.' Roy pointed out there was already a big project, partly for defence, to build a broad-gauge railway from Kandla to Bhatinda in Punjab, where 80 per cent of India's grain comes from. Giving Cargill a whole island would also have given it the most important creek, and with it control over one of the most important ports in India. If it was salt production that really interested Cargill, any other place would have been far better. 'I still wonder how they will come back', concluded Roy.[195]

Roy was looking at possible grain exports, but Cargill's position regarding the port had other foundations. India's consumption of fertilizer was expected to grow 17 per cent to 15 million tonnes in 1994. Although domestic production was expected to rise 15 per cent to 10.5 million tonnes, it was expected that about 7 million tonnes would have to be imported.[196] Cargill would obviously want to be a supplier of that fertilizer.

Cargill has patience, and it does little, if anything, rashly. If it cannot occupy the targeted region and a tactical retreat is required, it will regroup and try another manoeuvre. If a strategic partnership or a joint venture seems to be necessary to gain entry, Cargill will not let pride stand in its way. So it was hardly surprising that less than a year after it apparently left, Cargill was back in the Kandla area, making new efforts to establish a beachhead. In the last days of 1993, the Ahmedabad edition of *Indian Express* published a small news report stating that the Gujarat government had allotted 4,200 hectares in Mundra district to Adani Exports Pvt. Ltd and Adani Chemicals for production of industrial salt. The report did not mention that a public interest petition against this project had been filed in the Gujarat High Court when the people learned about the massive project. In response to the petition, the government insisted that no land had been allotted to anybody, except for 880 hectares granted to a company called Adinath Polyfils.[197]

The government response failed to satisfy the petitioners against the project, who pointed out that construction was already in progress on the Adani site, that the companies were in the process of building a road, and that they had awarded a contract for a massive private port to be built on 100 hectares of land allotted to them by the Gujarat Maritime Board. The Adanis had originally sought 6,400 hectares, according to the journal *Frontline*, but since 1,800 hectares of the coastal mangrove swamp had been designated as a central reserve forest, they had been granted only 4,200 hectares of 'coastal saline wasteland', despite the fact that much of this 'wasteland' is itself 500-year-old mangrove swamp. Construction of the jetty and roads made it impossible for about 4,000 fish workers to utilize the creeks as passage to the Gulf of Kutch where they traditionally fished. By the beginning of 1994 the Fisheries Department had already withdrawn the reef fishers' licences to fish in the sea as the coastal land had been acquired by the Adani companies. The Adani project includes full facilities for salt processing.

Cargill reappeared on the scene in April, 1994, in the form of a three-member team that included the company's Australian salt expert Richie Henry. The team expressed an interest in buying salt for the production of caustic soda, saying that they had come to the conclusion, after searching the world, that Kandla was 'the salt capital of the world' and that it would ideally suit the company's interests.[198] While the Cargill team did not say whether the company planned to set up the caustic soda plant in Kandla, local sources said that the local salt was not of high enough quality for caustic soda production and that therefore Cargill would have to build a plant there to process the salt as well as to manufacture caustic soda. Meanwhile, reported *Indian Express*, Cargill stressed that it had absolutely no interest in acquiring land or shipping facilities in the KPT area and that it was evaluating the possibility of buying salt from local producers. An experienced Gandhian organizer who visited the area reported, however, that after the Cargill team had visited Kandla, the people were reserved, somewhat hostile, and uninterested in Cargill's proposals, with the result that the Cargill team left without any concrete results.

Cargill is, nonetheless, patient, and in 1998 the company notified the public that:

> along with local partners, [it] has developed an anchorage lighterage facility* capable of discharging and loading Panamax vessels at the port of Rozy in the Gulf of Kutch along the northwest coast of India. The facility will have capacity to handle 800,000 metric tonnes of dry bulk commodities per year. Cargill will utilize the facility to import fertilizer and wheat as well as to export protein meal and other products ... Local partners will construct a 100 metre barge pier at the port of Rozy.[199]

Three years later, in 2001, Cargill reported that the first panamax vessel had completed discharging more than 55,000 metric tons of Cargill DAP from Tampa at Rozy and that a second panamax vessel would discharge at the facility a few months later. Cargill described its Rozy project as 'an integrated vessel discharge and loading facility, capable of discharging panamax vessels of fertilizer and wheat as well as loading panamax vessels of protein meal and other products'.

Cargill has a three-year agreement with the Punjab government for direct procurement of rice and wheat and is busy setting up elevators in the state of Punjab. It is also considering contracting directly with farmers for their crop, but it wants only the best quality, which Cargill will handle as an IP crop. Cargill also purchased a 'sick' roller flour mill at Noida and is upgrading it while carrying on talks with Punjab Agri Export Corporation about setting up an equally large integrated flour mill in Punjab. If Cargill succeeds in this, then 'almost all flour mills in Punjab will necessarily have to pull down their shutters as they would become unviable and not able to compete with the new entrant', according to the *Economic Times*.[200]

---

* An 'anchorage lighterage' facility is one where ships anchor a short way offshore and are loaded/unloaded by a crane barge onto barges for transfer to a shore-based facility. In Cargill's case, it means that their Indian partners are onshore while it remains a safe distance offshore.

San Francisco Bay

As the plane made a wide sweep over the South Bay before heading north into the San Francisco airport, I got a bird's eye view of Cargill's 11,600 hectares of salt flats that occupy almost all of the South Bay. Later I was taken on a guided tour (not by Cargill) of the mammoth salt ponds, which were all wetlands before they were gradually dyked and turned into salt-concentration and crystalliza-tion ponds from the middle of the nineteenth century. The ponds are currently producing about 1 million tons of salt a year with a value of about $20 per ton for the raw salt. If you wonder why you do not see Cargill salt in the supermarket, it is because it appears as Morton Salt. Some time after seeing Cargill's San Francisco salt operation I was visiting a trade show and came across a Morton Salt booth. I asked the woman in attendance what the connection there was between Morton and Cargill. 'None', she told me. I asked if Morton wasn't packing Cargill salt in the South Bay. She said, 'No, they have their own facilities.' 'But I saw the Morton plant next to Cargill salt operations', I said, to which she replied, 'I cannot comment on that', and smiled pleasantly.

The salt-making process begins when sea water is pumped from the tidal bay through a system of ponds for concentration. It remains in each pond for a period of time while solar energy evaporates water and concentrates the brine. Finally it is pumped to a 'pickle pond' and from there, at full saturation, into crystallizers. These are actually another form of pond made when the saturated brine is held in a basin with a compacted salt floor. In these basins the salt grows into crystals at which point the brine is drained off and the crystallized salt is 'harvested' in a process that is not unlike the removal of the top layer of asphalt off a highway before repaving, though in the case of salt the whole process is referred to as 'farming'. After this the salt is 'washed' to remove impurities and dried before being stockpiled for bulk shipping.

A variety of individuals and organizations are now trying to get at least 6,800 hectares of these salt ponds, which Cargill uses but does not own, restored as wetlands. (Cargill owns the other 4,800 hectares.) The Bay area environmentalists would really like to see all 11,600 hectares restored to their natural wetlands condition. They see it as simply outrageous that a private company should control such a large area of what once were public lands and marshes.

Cargill sold a 6-hectare parcel on San Francisco Bay in Redwood City – a local landmark for decades with its huge salt pile – to pharmaceutical giant Abbott Laboratories in 2000. Earlier in the year Cargill had sold more than 128 hectares of salt pond property to the Santa Clara Valley Water District. At the end of 2000, it was reported that Cargill was still on track to sell about 7,600 hectares of salt ponds to state and federal agencies for restoration. Cargill would still operate on about 4,800 hectares on leased land in the South Bay.[201] The whole business remains unsettled as Cargill seeks to maximize the price it might accept and the state and environmental organizations push to have the whole area restored. (There is also the complicating issue of possible airport expansion through filling some of the Bay, which would only be possible if there was some mitigation, such as reclaiming Cargill-controlled salt marsh.)*

What happens to old salt ponds? In 1994 Cargill sold an area one-third the size of the city of San Francisco to the California Wildlife Conservation Board for $10 million, one-third of its appraised value of $34 million. The $25 million difference was to be 'donated' to the state, making it the company's largest single dedicated environmental contribution ever (worth how much in tax reduction?). From an ecological perspective, the area is valued as a mammoth wetland or salt marsh. The 4,000-hectare tract north of San Francisco was used in the production of solar salt from the 1950s to 1990 when the company lost its main customer, Dow Chemical Co., following the closure of its plant there.

Venezuela

Note: I have not travelled to Venezuela and the following account is based on information provided by local researchers who have worked for many years with the inhabitants of the region of Zulia state discussed here.

Cargill explains on its Venezuelan website: 'By the end of the '80s, the shortage of salt for industrial purposes in Venezuela induced the private sector and the Venezuelan government to join forces to develop a solar salt project, which crystallized with the incorporation in 1989 of Productora de Sal, CA (Produsal). Today, the shareholders of the company are Petroquémica de Venezuela, S. A. (Pequiven), 30 per cent, and Cargill de Venezuela, 70 per cent.'**[202]

---

*   Late note: At the end of May, 2002, it was announced that Cargill had agreed to sell 6,600 hectares of salt ponds around the south end of San Francisco Bay. The $100 million price agreed upon is to come from federal, state and private foundation sources.

**   Nowhere does Cargill describe the coastal environment in which Produsal operates or the human communities affected by it other than saying that the project 'generated' 3,000 jobs in the construction phase.

Produsal obtained a concession (registered in the Gazette of Zulia State, March 1993) to produce solar (evaporated) salt for a period of 50 years subject to renewal within Los Olivitos Swamp wildlife and fishing reserve. Construction started in 1994, ended by 1998, and went into commercial operation in January 1999 with a capacity of 800,000 tons per year from the 5,000 hectares of evaporation ponds constructed in the Los Olivitos marsh. According to Cargill, the plant has become the main supplier for Venezuelan industrial use, human consumption and animal feed. Pequiven is the main user of industrial salt in Venezuela as a raw material to manufacture chlorine and caustic soda. The oil industry also uses salt as a component of oil and gas drilling fluids. Chlorine's main use is in the manufacturing of PVC. It is also used for the treatment of potable water.

The Los Olivitos Marsh is a coastal wetland of 33,000 hectares of mangrove swamps, salt marshes, sandy beaches, and dunes lying within the Maracaibo estuary, located between the Gulf of Venezuela (Southern Caribbean Sea) and Lake Maracaibo. Of the Olivitos estuary 15,000 hectares was declared a Wildlife Refuge and Fishing Reserve under Venezuelan law and in 1996 this wetland was listed as a Wetland of International Importance according to the United Nations' Ramsar Treaty.

The marsh receives the waters of El Tablazo Bay on the west and the waters of the Caribbean to the north. It is fed by two freshwater rivers and is an important resting, feeding and nesting place for many species of birds, as well as being a nursery zone for several commercial fish species, crustaceans, and other aquatic organisms. The mangrove areas of Los Olivitos and San Carlos supply 50 per cent of the catch of Zulia state, most of it coming from the artisan fishery. Zulia state actually exports white shrimps and blue crabs to the US.

El Tablazo Bay lies north-east of Maracaibo City and is now considered part of Lake Maracaibo. Originally, Lake Maracaibo was almost entirely closed to the sea by a bar which provided an effective obstacle to big oil-tankers entering the lake, but this obstacle was overcome by dredging out the bar. The continuous dredging is one of the most important pollution sources in this area due both to the movement of sediment and the increasing salinity of the lake itself. In the past the lake was not salty because the rivers brought enough fresh water in to keep the sea water out. In its natural state the lake would not have been suitable for industrial salt production.

Through Produsal, Cargill intends to supply all the salt required by the El Tablazo Petrochemical Complex, where Petroquimica De

Venezuela uses salt as a primary feedstock for chlorine production and the manufacture of PVC. Large amounts of salt are also used in the nearby oil-field operations in Lake Maracaibo and its vicinity. Contrary to Cargill's description, the amount of salt used in chlorine production for domestic water purification is a very small percentage of the total.

Since 1995 the 1,700 villagers of Ancón De Iturre have been resisting Cargill's efforts to develop salt production in their home. The community, prior to the arrival of Produsal, was a productive village with minimal unemployment thanks to its way of life characterized by traditional forms of labour such as fishing and artisanal production of salt. The imposition of the labour-saving salt ponds left the community without the employment which artisanal production of salt provided and approximately 300 inhabitants of the village were rendered redundant. Not only was almost the entire active population of Ancón de Iturre without work, but between 1,500 and 1,700 labourers from the communities of Boca del Palmar and Quisiro lost their means of livelihood. In addition, the artisanal fishery has been severely reduced.

However, an even greater threat arose in 1999 with the plan of Produsal to install a pipeline to discharge bittern (*amaragos* in Spanish), a highly alkaline toxic by-product of salt production, directly into Lake Maracaibo, even though Cargill knows very well the environmental damage that will cause. Six years before Cargill acquired Leslie Salt's San Francisco and Australian solar salt operations in 1978, Leslie had commissioned a scientific report on its proposed discharge of bittern into San Francisco Bay. The report describes the toxicity of bittern and points out that salt production by solar evaporation produces one tonne of bittern for each ton of salt produced. The same report indicates that bittern must be diluted by at least 100 parts to one with fresh water before losing lethality. Current environment regulations in San Francisco Bay require that Cargill Salt dilute its bittern discharge at least 300 parts to one and then release it only during an extra strong ebb tide and at locations where there will be strong mixing and tidal dispersion. These conditions are so stringent that bittern is currently not discharged but is stored in dyked bittern ponds.

In 1999, when Produsal attempted to install a bittern pipeline, the men, women and children of all the fishing families of Los Olivitos stopped the installation by placing themselves in the way of the construction machinery. Fortunately no one was injured and they halted construction of the pipeline. At that point Produsal retreated and the Environmental Ministry cancelled Produsal's permit and assured the villagers that a public hearing would be held before a new permit was issued.

The promised hearing never did take place, but all remained quiet until just before Christmas 1999 when Produsal crews reappeared with a new pipeline permit and again laid out tubing. Feeling betrayed, the angry villagers of Ancón de Iturre, now joined by residents of the neighbouring villages, formed a 1,000-strong protest group and demanded both a meeting with Produsal representatives and that the pipes be taken back to salt company property. The peaceful protest took violent form when, after the multitude demanded the dismantling of the pipeline, Produsal managers responded with jeers while a truck with armed personnel burst into the crowd. The armed guards began to shoot and immediately the crowd reacted by burning the pipe and the truck. Produsal accused some fishermen and local leaders of causing damage to its properties and initiated a judicial process.

As one might expect, when dealing with a very large TNC and its local partners, there is more to the story. The bittern discharge was quite literally just the end of a very long 'pipeline'.

Venezuela President Chavez's development administration wants the El Tablazo petrochemical complex to be the focal point for global production of PVC. PVC production requires vast amounts of chlorine which comes from salt which, in the case of El Tablazo, is made by Produsal in the Los Olivitos marshes.

It is this strategic position of salt as a raw material to produce PVC, and also as an additive in mud for oil drilling, that drove the Venezuelan state in 1968 to reorganize the state monopoly of salt and transfer it to the Empresa Nacional de Salinas. In 1995, under free-market liberal policies and privatization, Cargill bought out Grupo Zuliano's participation in Produsal.

Roskill Consulting Group pointed out in a study in 2000 that because of the pollution it causes, PVC production is being banned in industrialized countries and shifted to third-world countries, opening up a very big market for solar salt which, as we have seen, is also being pushed out of North America for environmental reasons. However, as we have also seen in the cases of India and Venezuela, which are distressingly similar examples, Cargill's salt imperialism is not necessarily welcomed with open arms by the inhabitants of the regions it would invade.

19   Only Cargill's Future?

One could be tempted to describe Cargill's rapid repositioning over the last five to seven years as radical, but that would be doing the company a disservice. Cargill brings its history with it and there is a clear logic to the changes it has made. These changes have been of the company's own choosing in the interest of its own immortality.

Reflecting on the main character in the text I have written, I see Cargill, more limber than ever despite its age, building on the strengths of its long experience, but not captive of them, to reposition itself, taking advantage, as it does so, of the mistakes, bad judgement, limited vision, wishful thinking and big egos of its less fit competitors, suppliers and customers. In recent years Cargill has forsaken a number of enterprises, such as fresh fruit, rubber and coffee trading, hybrid seeds, equipment leasing and transportation services, while finding new ways to extend the product lines of its traditional businesses, such as soy and corn milling. And then there is always the complex but firm foundation of its financial services and financial markets activities – good old trading, speculating and 'risk management'. Like a healthy organism, Cargill's old cells constantly die and new ones take their place.

Most interesting, to my mind, is what Cargill has been up to in the creation of joint ventures and partnerships in the business activities it is most familiar with. There is an aspect of this partnering that is very disturbing. A great many of Cargill's new joint ventures are with farmers' cooperatives, from small single-facility grain coops to very large coop conglomerates such as CHS. I've already described Cargill's new flour-milling venture with CNS. Here is a sampling of its recently formed smaller joint ventures.

In 1997, Cargill's structural transactions included the acquisition of 20 grain elevators in central US, the construction of at least one new grain elevator, and the upgrading of others. Cargill leased its York, North Dakota, elevator to BTR Farmers Co-operative and BTR became a 'preferred supplier' of grains and oilseeds to Cargill. It formed a limited liability company with Garden City Coop to own and operate the grain handling facility that Cargill was expanding in Garden City, Kansas. Garden City Coop is an 80-year-old grain marketing and farm supply cooperative with 1,500 members. Cargill formed another joint venture with Alceco, a farmer-owned cooperative in Iowa, to combine the grain-handling, fertilizer, agrotoxin, seed and feed operations of the two companies. In Indiana Cargill purchased Frick Services' four grain facilities and the two shortline railways serving them as well as Heartland Cooperative's grain elevators in eastern Illinois and the nine elevators of AGP Grain in Indiana and Ohio. In Kansas Cargill and Santana Cooperative made a deal in which Santana purchased five Cargill grain elevators and Cargill agreed to market the grain Santana collected.

In 1997 Cargill also took over operation of the corn wet-milling plant owned by ProGold Limited Liability Co. in Wahpeton, North Dakota. The plant, built by a consortium of three farmers' coops (Golden Growers Coop, American Crystal Sugar Co. and Minn-Dak Farmers Coop), came on stream the year before but it was on the brink of bankruptcy due to poor market conditions when Cargill began to operate it on a ten-year lease. Early in 2001 Cargill announced that it was closing down operations at ProGold, citing rising energy costs and poor market conditions for its decision, though Cargill said it would continue to make lease payments to the coops. It subsequently reopened the plant.

That was all in one year, and no doubt there was more that went unreported. In every case, what might have looked like a good marketing opportunity for a small farmer-owned cooperative was, in fact, an arrangement that assured Cargill a reliable supplier of grains and oilseeds without increased investment. However they might like to think of it, the farmers become captive suppliers to Cargill.

While the coops may appear to still be there, from Santana to CHS, the fact is that they have been effectively absorbed, leaving Cargill the beneficiary of a century of dedication and hard work of earlier generations of farmers who were building lives, and businesses, for themselves they thought. In Canada there are the sad examples of the big prairie grain pools (cooperatives) which have become capitalist enterprises serving their shareholders, not their (mostly former) farmer members. Alberta Pool and Manitoba Pool merged to form Agricore, which was then taken over (they called it a merger) by United Grain Growers (UGG) to form Agricore United. UGG was the first of the prairie coops to go capitalist with a public share offering several years ago, which gave ADM control with something like 42 per cent of the shares. The 'merger' of Agricore with UGG meant that what had been the Alberta and Manitoba coops also became capitalist since they had to be folded into UGG's share structure. Then ADM appointed its two top executives as members of the board of Agricore United, indicating that ADM intended to exert its power. The remaining big coop, Saskatchewan Wheat Pool, allowed management to take over and with visions of becoming a transnational conglomerate dug itself deep into debt with acquisitions. It has spent the past two years selling itself off piece by piece to reduce its debt and to keep its public shareholders at bay – it had followed UGG into the public capital market to finance its acquisitions.

> Prairie farmers who are watching the Pool die its death by a thousand cuts must be filled with conflicting emotions. Anger is no doubt the predominant one, but there must also be a hint of sadness and regret for many. Regret that the prairie dream of controlling their economic future by controlling their 'merchant of grain' must now be given up. Regret that the company built and so carefully nurtured by fathers and grandfathers should be undone by the folly of a few short years. Regret that, seeing it happen, farmers were so powerless or unwilling to prevent it. And more than a few farmers must be wondering how a company could be so badly managed as to be in this position.[203]

Strategic partnerships don't only occur in traditional businesses. They now include universities as they take on more and more corporate characteristics. Kent State University in Ohio, for example, has become a 'core school' in Cargill's Higher Education Initiative, a corporate programme designed to promote strategic business partnerships with several colleges and universities in the US. In Kent State's case, this translates into $300,000 over three years for the university's college of agriculture to help students and faculty develop a better understanding of today's agribusiness sector (i.e. Cargill) and to enable the university to better serve the agriculture industry, according to the university's dean of agriculture.[204]

While not a partnership in name, the gift of $10 million in 1999 – the largest single gift in the company's history – to the University of Minnesota as half the cost of a new building (the gift has to be matched by the state) to be devoted to the decoding of microbial and plant genomes might not give Cargill any short-term returns, but the possibilities of long-term benefits and nurturing the culture of university–corporate collaboration must have occurred to the Cargill directors.

Thinking of Cargill as an immortal organism, one could view the storm of acquisitions, divestitures, joint ventures and partnerships that Cargill has been engaged in for the past five to seven years as a recognition of the importance of biological diversity.

I have already indicated that Cargill is ecologically minded and environmentally sensitive, but there are accidents and mistakes and these usually raise questions, for some of us, about the nature of the operations themselves. It is not enough, for example, for Cargill Pork Inc. to pay a $1 million fine and $51,000 in restitution for the illegal dumping of pig waste that contaminated 8 km of a central Missouri river and to spend $500,000 in remediation costs associated with the dumping as well and then say, 'We're pleased to settle the matter and are satisfied with the terms of the settlement ... The incident clearly concerned us and was not characteristic of Cargill Pork's environmental record. We can now move ahead and put this matter behind us.'[205] It is not enough to 'put this matter behind us'. The question remains, why was a facility that had the potential to cause such pollution built, or allowed to be built, in the first place?

Cargill can legitimately take pride in its phosphate-mine site restoration in Florida and in its solution to disposal of the waste water from its Alberta meat plant. But to describe Cargill as a good ecological citizen on the basis of such individual cases would be to miss the larger issues altogether. The mining of huge amounts of phosphate rock in one location to produce fertilizer that is then shipped around the world is not ecologically sound. The concentration of great numbers of cattle in one area so that it is possible for Cargill to kill thousands of cattle in one day in one place, day after day, is neither environmentally nor ecologically good practice. For Cargill to maintain that it is doing Indian farmers a favour by offering them hybrid sunflower seed that is composed entirely of alien germplasm is the antithesis of sound ecology.

The creation of dependency is an ancient colonial practice, serving the interests of the colonizers at the expense of the colonized. I have elsewhere likened hybrid seed to an envelope within which is contained its relations of production (see Kneen, *The Rape of Canola*). Looking at Cargill's activities in India, it is not hard to imagine seed in the role of colonizing troops, the occupiers of the land dictating that the peasants will now produce agricultural commodities for the colonial power, which will take these commodities (perhaps to another land), process them, and send them back to be purchased by those among the colonized peoples who can afford them. This is exactly what the British did to the textile industry in India; it is what Gandhi protested against, and it is what Cargill would have reproduced with its hybrid sunflower and corn seed – at the same time as it would be creating customers for its fertilizers. But Cargill apparently concluded that it had better things to do than provide the forward troops that would draw the enemy fire. It sold its seed business to Monsanto enabling it to concentrate on being the crucial supplier to the troops and the occupying force after the initial battles have been won.

The global process in which Cargill is engaged can also be described as the recreation of feudalism, with the intent of driving people off the land by what amounts to acts of enclosure, forcing them to become wage labour and customers for what they used to provide for themselves. This is the process which still goes under the misleading name of development.

Current corporate – and to a great extent now public – ideology holds that the corporation is the fount of wisdom and the most competent body to plan global production and distribution in accordance with the dictates, or ideology, of the market. Accordingly, Cargill now puts itself forward as the most competent agency to help develop the backward (that is, unindustrialized) peoples of the world. At the same time, these same companies are heavy feeders at the public trough, while, with their mouths full, they decry public indebtedness and social welfare. This suggests to me that their business success may at times have more to do with their ability to avail themselves of public subsidies than with their business acumen. Cargill is no exception.

Cargill's corporate goal was stated, at one time, to be the doubling of its size every five to seven years, and while it may have relieved itself of the burden of such statements, the achievement of such a goal requires the occupation of ever more territory and the expulsion of whole societies from their settlements and their commons. Cargill emphatically proclaims that in the long term this will be beneficial, since the outcome will be a higher standard of living as these people will be able to buy a greater variety of food at lower cost than they could produce for themselves. No system of subsistence agriculture can ever achieve such benefits, it says, assuming that everyone will somehow have the money required to purchase what they need and what Cargill is willing to supply.

Cargill's argument is not, of course, a matter of science. It is a question of ideology, or faith, because there is no proof or even anecdotal evidence that the outcome would ever be as Cargill predicts. So we come back to the thesis of this study: Cargill does not really do business in food. It deals in agricultural commodities as raw materials to be deconstructed and reconstructed into some value-added product for the market in order to produce a profit for the corporation. It does this with consummate skill.

Cargill and the advocates of science and technology, progress and capitalism, claim that theirs is the only way forward and the only hope for feeding an expanding global population. We must remember, however, that the globalized industrial system that works for Cargill is a very recent invention – post-1945 – that has worked well to make Cargill and a small elite of the world wealthy, but at an increasingly unacceptable cost to the earth, to the creatures of the earth, and to the majority of the people of the world. The industrial system may be able to produce quantities of food, but it cannot produce the justice required to ensure that everyone is adequately nourished.

I cannot contain or control Cargill as the *zaibatsu* of Japan may be able to, and my influence over the World Bank or the WTO is rather less than Cargill's to say the least. On the other hand, there are many things that Cargill cannot do and many things that Cargill does not want to do. Its structure and business are contradictory to decentralization and self-provisioning. Cargill deals in volume, and to get sufficient volume in both buying and selling it has to do business transnationally and industrially. In other words, it is a matter of both scale and mode of operation, and there is a definite threshold beneath which a company like Cargill cannot function even if it wanted to. Therein lies the key to resistance and the pursuit of alternatives.

The Japanese *zaibatsu*, and to a lesser extent the Korean *chaebols*, practised one kind of resistance to Cargill, banding together as warlords to defend their territory. The farmers of India, in their numbers, have manifested a very different kind of resistance to Cargill's attempted invasion, while the outnumbered small-scale farmers of Japan, Cuba and many other countries are practising a parallel strategy of resistance: small-scale diversified agriculture and the development of local self-provisioning food systems – a recreation of the commons.

The choice before us can be put in terms of the deepening divergence between hybrid structures and organizations and the practice of monoculture on the one hand and open-pollinated organizations and the practice of diversity on the other. The metaphor, of course, refers to fundamental differences in seed characteristics and propagation and in the cultures of their production and reproduction.

Modern hybrid seeds produce deliberately uniform commodities as the foundation of industrial agriculture. They are not themselves capable of reliable self-reproduction but are, instead, dependent on an external industrial process for their replication. In contrast, traditional seeds are, by nature's necessity, open-pollinated and self-replicating, not dependent on outside powers (unless you count the sun, the wind and the birds and bees) and will themselves generate cultural diversity through mutation and cross-breeding.

Genetically engineered seed (herbicide-tolerant soy, corn and canola, or Bt corn and cotton) also follows nature's inherent drive to reproduce and in a sense rides on this to contaminate the countryside and traditional seeds, and, while this contamination may be the deliberate sabotage of nature's diversity by those such as Monsanto, nature is quick to respond with mutations and adaptations that produce biological mechanisms to overcome such assaults. Therein lies our hope.

Cargill and other TNCs have the wealth, skill and political leverage to outflank or overpower virtually any head-on attacker, and the game is rigged in their favour. They cannot, however, force people – either farmers or the general public – to play the game.

The refusal to use hybrid or patented seed (or highly processed food that has travelled from some centralized production facility) and the rejection of industrial monoculture (franchised fast-food) is the beginning of resistance. The deliberate use of traditional open-pollinated seed (figuratively and literally) and the pursuit of diversity and self-reliance are the basis for building ecologically sound and socially just alternatives.

Around these old affirmations and new beginnings a new genus of 'open-pollinated' social organization is emerging: communities that thrive on, and in turn generate, diversity and inclusivity. They share a recognition of the interdependence of every organism and the identification of personal long-term well-being with the good of their community and of society as a whole.

It's hard to imagine a place for Cargill, or any other food transnational, in such a community.

Notes

Dates shown here are in short form 21/8/01, meaning 21 August 2001.
Abbreviations used in the notes.

CB – *Cargill Bulletin* (no longer published)
CN – *Cargill News* (monthly publication for company employees)
CRM – *Corporate Report Minnesota*
G&M – *Globe and Mail*, Toronto
M&B – *Milling & Baking News*
M&P – *Meat & Poultry* magazine
MC – *Manitoba Co-Operator*, Winnipeg
NYT – *New York Times*
ST – *Star Tribune*, Minneapolis
WGB – W. G. Broehl Jr, *Cargill – Trading the World's Grain*, University Press of New England, New Hampshire, USA, 1992
WSJ – *Wall Street Journal*

1. www.cargill.com, 20/2/02.
2. *M&B*, 21/8/01.
3. Cargill executive Peter Kooi in *M&B*, 17/11/98.
4. *M&P*, 4/01.
5. *M&B*, 2/6/98.
6. *M&B*, 16/10/01.
7. *M&B*, 16/10/01.
8. *M&B*, 27/6/00.
9. *M&B*, 23/10/01.
10. Phone interview, Jim Snyder, Dun & Bradstreet, 10/10/94.
11. Wilson, J.R., 'A Private Sector Approach to Agricultural Development' manuscript, Cargill Technical Services Ltd, UK, 1994.
12. Remarks by Whitney MacMillan before the Columbus [Ohio] Council on World Affairs, 15/12/92.
13. VP Robbin Johnson to the USDA Outlook '93 Conference, *M&B*, 22/12/93.
14. MacMillan 15/12/92.
15. *Asia Pacific Economic Review*, Summer/Autumn 1996.
16. *CB*, 2/98.
17. *ST*, 8/7/98.
18. Cargill Internal Memo, 18/6/99.
19. Cargill Internal Memo, 19/7/99.
20. Cargill VP Jim Prokopanko, Sioux Falls, South Dakota, 20/10/99.
21. Bob Parmelee, President, Food System Design, 25/6/01.
22. *ST*, 6/5/94; *Forbes*, 5/12/94.
23. *ST*,18/2/94.
24. *WSJ*, 9/1/97.

25. *ST*, 6/2/98,17/4/99.
26. Cargill press release,15/1/02.
27. W. Duncan MacMillan, with Patricia Condon Johnson, *MacGhillemhaoil – An Account of My Family from Earliest Times*, privately printed at Wayzata, Minnesota, 1990 (two volumes, illustrated).
28. WGB, p. 686.
29. *CRM*, 1/93.
30. *CN*, 10/91.
31. *M&B*, 11/2/93.
32. Archer Daniels Midland annual report 1994.
33. *M&B*, 11/2/93.
34. *CN*, 11/91.
35. *M&B*, 11/2/93.
36. *M&B*, 11/2/93.
37. *ST*, 18/5/86.
38. *Fortune*, 13/7/92.
39. *ST*, 29/6/93.
40. *Fortune*, 28/6/93.
41. *ST*, 17/4/99, 7/6/99.
42. *Dyergram*, 21/3/01.
43. *M&B*, 7/12/93.
44. *CN*, 11/93.
45. *CN*, 2/93.
46. *CN*, 2/93.
47. *CN*, 11/93.
48. *ST*, 29/6/93.
49. *CN*, 6/93.
50. *CB*, 10/88.
51. Family Farm Organizing Resource Centre, St Paul, n.d.
52. Richard Gilmore, *A Poor Harvest*, Longman, 1982, p. 138.
53. Ralph Nader & Wm Taylor, *The Big Boys*, Pantheon, 1986, pp. 322–3.
54. WGB, p. 778.
55. *G&M*, 5/12/86.
56. *M&B*, 28/11/89.
57. *NYT*, 10/10/93, 1st of three articles; 10, 11 & 12/10/93, by Dean Baquet with Diana Henriques.
58. *M&B*, 14/11/89.
59. *M&B*, 17/8/94.
60. Cargill press release, 6/11/01.
61. *Cattle Buyers Weekly*, 26/9/94.
62. *Ontario Farmer*, 16/11/88.
63. *Farm to Market Review*, 7/93.
64. Cargill press release, www.cargill.com
65. Canadian Press, 21/590.
66. *Farm & Country*, Toronto, 21/11/93.
67. *Financial Times*, Canada, 13/5/91.
68. *CN*, 2/93.
69. *ST*, 15/7/95.
70. Cargill News International, 1999.

71. Reuters, 28/9/99.
72. *CRM*, 8/85.
73. Ibid.
74. Ibid.
75. Ibid.
76. www.cargill.com, 26/9/97.
77. Ibid.
78. *G&M/WSJ*, 30/9/97.
79. *M&B*, 20/4/99.
80. *WSJ*, 29/12/95.
81. *M&P*, 3/94.
82. Personal interview, 28/2/94.
83. *CN*, 12/90.
84. *ST*, 20/7/93.
85. *CB*, 10/94.
86. Cargill Update, Winter 1994.
87. *Far Eastern Economic Review*, 27/10/94.
88. USIA,11/8/97.
89. *Fortune*, 13/7/92.
90. www.cargill.com, 3/4/97.
91. Cargill press release, 10/4/96.
92. *ST*, 20/9/98.
93. www.cargill.ven, updated 8/00.
94. *M&B*, 30/1/96.
95. Cargill press release, 8/6/98.
96. *WSJ*, 31/10/01.
97. Speech to the Corn Refiners Association, 2000.
98. *DowJones News*, 30/10/01, 31/10/01.
99. *M&B*, 29/1/02.
100. Pat Thiessen, quoted by David Fry, assistant administrator for the Kansas Wheat Commission, in MC, 30/3/95.
101. *M&B*, 25/9/01.
102. www.admworld.com
103. *Forbes*, 18/9/78.
104. Cargill brochure, Ontario, 1989.
105. *Fortune*, 25/7/94.
106. Ibid.
107. *M&B*, 1/11/94.
108. WGB, pp. 772–4.
109. David Rogers, president, Financial Markets Division, *CN*, 1/94.
110. *CN*, 1/94.
111. Cargill Update, Winter 1994, and corporate brochure, nd.
112. Cargill brochure, nd.
113. *ST*, 28/2/95.
114. *ST*, 31/10/95.
115. *ST*, 11/11/96.
116. *G&M*, 22/8/97.
117. *ST*, 23/12/97.
118. *ST*, 16/5/98.

119. *ST*, 2/10/98; GM, 21/10/98.
120. *ST*, 28/12/01.
121. ccc.cargill.com/fmg/
122. Kevin Phillips, *Arrogant Capital*, Little Brown, 1994, pp. 79–80.
123. St Paul *Pioneer Press*, 24/9/01.
124. Cargill, press release, 4/4/00, www.cargill.com
125. *CB*, 11/88.
126. WGB, p. 554.
127. *CN*, 2/92.
128. *El Financiero International*, 19–25/7/93.
129. *ST*, 7/12/93.
130. *M&B*, 10/8/99.
131. WGB, p. 722.
132. www.cargill.com, updated 8/00, accessed 15/2/02.
133. Cargill corporate brochure, 2001.
134. 'Soybean Cultivation as a Threat to the Environment in Brazil', Philip M. Fearnside, Department of Ecology National Institute for Research in the Amazon, Manaus, Amazonas, 3/10/00.
135. *M&B*, 8/1/02.
136. Fearnside, 3/10/00.
137. David Kaimowitz and Joyotee Smith, 'Soybean technology and the loss of natural vegetation in Brazil and Bolivia', in *Agricultural Technologies and Tropical Deforestation*, A. Angelsen and D. Kaimowitz (eds). CAB International, Wallingford, UK. 2001, pp. 195–211.
138. *Financial Times*, 20/11/96.
139. Glenn Switkes, 'Competition between Brazilian, U.S. growers needs unmasking', *Feedstuffs*, 30/4/01.
140. www.aclines.com, accessed 15/12/01.
141. Fearnside, 3/10/00.
142. *Journal of Commerce*, 3/1/96.
143. Fearnside, op. cit.
144. *CN*, 6/93.
145. *CN*, 6/93.
146. *Dyergram*, 29/11/01.
147. *International Bulk Journal*, 4/92.
148. *Herald Tribune*, 2/9/87, 25/9/87.
149. *CN*, 5/94.
150. Anthony Depalma with Simon Romero, *NYT*, 24/4/00.
151. *Packer*, 10/7/92.
152. *Packer*, 18/12/93.
153. *Packer*, 2/9/96.
154. *M&B*, 15/3/94.
155. *Journal of Commerce*, 21/11/91.
156. *Activity News*, National Council of Churches in Korea, May–July 1990.
157. *Han-kyoreh Shinmun*, 24/8/89, translation.
158. *Korea Times*, 7/1/88.
159. Personal interview, 1/8/94.
160. Charles Alexander, personal interview, 1/8/94.
161. *CN*, 5/98.

162. Takashi Suetsune, *Journal of Japanese Trade & Industry*, #4, 1988.
163. Editorial, *M&B*, 22/3/94.
164. *Business Week*, 11/7/94.
165. 'Discover CNAL' (Cargill North Asia Ltd) no date.
166. Company transcript, 24/8/94.
167. Reuter European Business Report, 13/10/92.
168. *Nikkei Weekly*, 23/12/96.
169. *ST*, 23/12/96.
170. http://www.farmchina.com/clientwebsite/cargill, accessed 8/2/02.
171. http://english.peopledaily.com, 11/5/01.
172. www.cargill.com, under 'speeches'.
173. IPS – Interpress Third World News Agency, 4/2/97.
174. WGB, p. 746.
175. WGB, p. 749.
176. *CN*, 11/91.
177. Reuters, 2/2/99.
178. Cargill/Pioneer press release, 16/5/00.
179. Ibid.
180. Cargill press release, 14/5/98.
181. www.cargillhft.com
182. *The Other Side*, 11/93.
183. *Biotechnology & Development Monitor* #19, 6/94.
184. Cargill Seeds press release, 17/7/93.
185. *Times of India*, Bangalore, 30/12/92.
186. *India Express*, 15/8/93.
187. Personal interview, 1/2/94.
188. Personal interview, 12/1/94.
189. *Biotechnology & Development Monitor* #3, June, 1990; *Biotechnology & Development Monitor* #6, March, 1991, from Robert Walgate, 'Miracle or Menace? Biotechnology and the Third World', Panos Institute, 1990.
190. www.cargill.com, accessed 26/9/97.
191. San Francisco *Chronicle*, 13/3/01.
192. *El Universal*, 14/10/01; 7/8/00; www.cargill.ven
193. *Bloomberg News*, 15/8/01.
194. *Financial Times*, 7/5/93.
195. Personal interview, 14/1/94.
196. *Western Producer*, 13/10/94.
197. *Frontline*, India, 17/6/94.
198. *Indian Express*, Ahmedabad, 28/4/94.
199. www.cargill.com 10/3/98.
200. *Economic Times*, Ahmedabad edition, 15,24/7/99.
201. San Francisco *Chronicle*, 11/12/00.
202. www.cargill.com.ve, accessed 13/10/01.
203. Paul Beingessner, weekly column (e-mail) 24/2/02.
204. *Feedstuffs*, 22/9/97.
205. AP, 20/2/02.

References

Periodicals

Biotechnology & Development Monitor, Amsterdam, quarterly
Bloomberg News
Business Week
Cargill Bulletin
Cargill News
Cargill Publications – the term used by Cargill for items that are often undated and drawn from unnamed company sources
Cattle Buyers Weekly, Petaluma, California
Corporate Report Minnesota, Minneapolis, Minnesota, monthly
DowJones News
Dyergram, B.W.Dyer & Co., New Jersey, bwdyer@worldnet.att.net
Economic Times, India, daily
El Financiero International, Mexico City, weekly
Feedstuffs, USA, weekly
Financial Times, London, daily
FOLicht
Forbes
Globe and Mail, Toronto, daily
Grain & Milling Annual, Milling & Baking News, Marriam, Kansas
India Express, daily
International Bulk Journal, UK, monthly
Japan Agrinfo Newsletter – Japan International Agriculture Council
Japan Economic Journal
Manitoba Co-Operator, Winnipeg, weekly
Meat & Poultry, USA, monthly
Milling & Baking News, Marriam, Kansas, weekly
Mining Annual Review
Nikkei Weekly, Japan
Oils & Fats International, UK, quarterly
Ontario Farmer, London, Ontario, Canada, weekly
Post-Intelligencer, Seattle
Seattle Times
Seed World, USA, monthly
Star Tribune, Minneapolis, Minnesota, daily
Wall Street Journal, daily
Washington Post, daily
Western Producer, Saskatoon, Saskatchewan, Canada, weekly (WP)

Additional Works

William Cronon, *Nature's Metropolis – Chicago and the Great West*, Norton, 1991

A.V. Krebs, *The Corporate Reapers*, Essential Books, 1992 (Box 19405, Washington DC 20036 USA)

Patrick McCully, *Silenced Rivers: The Ecology and Politics of Large Dams*, International Rivers Network/ZedPress 1996, new edition 2001

Dan Morgan, *Merchants of Grain*, Viking Press, 1979; Penguin, 1980

Marc Reisner, *Cadillac Desert – The American West and its Disappearing Water*, Penguin, 1986, new edition 1997

Index

Compiled by Auriol Griffith-Jones